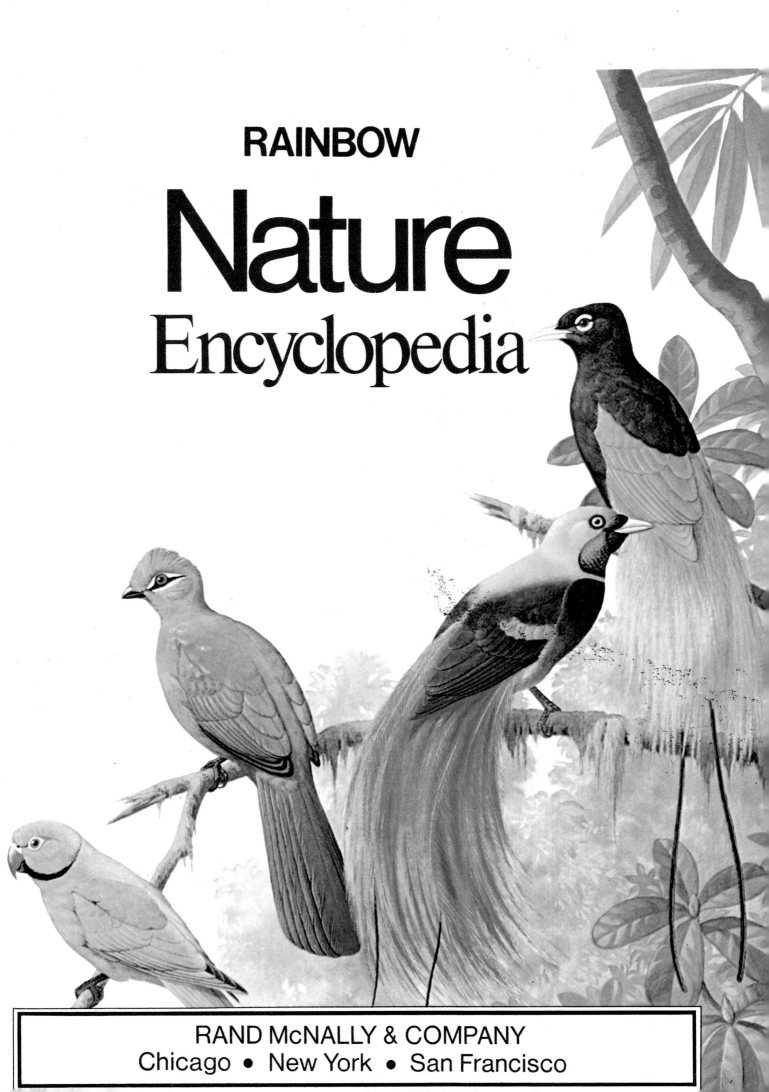

RAINBOW
Nature
Encyclopedia

RAND McNALLY & COMPANY
Chicago • New York • San Francisco

Contents

EDITORIAL
John Paton
Catherine Dell
Cover: Keith Groome

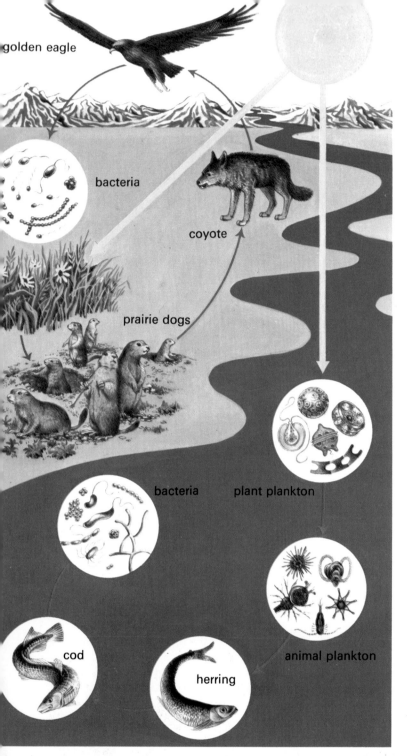

golden eagle

bacteria

coyote

prairie dogs

bacteria

plant plankton

cod

herring

animal plankton

All Kinds of Animals

Hundreds of millions of years ago, simple, microscopic creatures lived in the oceans. From these creatures came the many different animals alive today.

All animals are divided into groups according to their kind. To begin with, all animals are either protozoans or metazoans. Protozoans are tiny creatures with only one cell. (Cells are the minute building bricks that make up all living things, both animals and plants). There are over 30,000 kinds of protozoans. Most are too small to be seen without a microscope.

Metazoans are animals with many cells. Nearly all the animals man knows, from earwigs to elephants, are metazoans.

Many metazoans do not have a backbone. They are invertebrates. Invertebrates include the coelenterates, or hollowed-gutted animals, like sea anemones and jellyfish. The flatworms, roundworms and worms in segments are three more groups of invertebrates. Molluscs make up another group. Molluscs include slugs, snails, oysters and octopuses.

Many invertebrates, like ants, spiders and crabs, have jointed legs. These are arthropods. The last main group of invertebrates, the echinoderms, contains spiny-skinned animals like sea urchins and starfish.

Below: Animals at an African waterhole. Zebras, antelopes, guineafowl and giraffes are all plant-eaters. They can share the same patch of land because each kind of animal eats a different plant or a different part of the same plant.

Above: Food chains on land and sea. Each chain contains eaters and eaten. On land, prairie dogs eat plants. A coyote eats prairie dogs. An eagle eats a sick coyote. Tiny bacteria feed on the eagle when it dies and produce chemicals that nourish plants. In the sea, microscopic plants form food for tiny animals. These animals are eaten by herring. Cod often eat herring. At the end of the chain, bacteria feed on dead fish. On land and in the sea, all animals need plants. In turn, all green plants need sunlight to help them make food.

Metazoans that have some kind of backbone are called chordates. Chordates with true backbones and skulls are known as vertebrates. There are five groups of vertebrates: fishes, amphibians, reptiles, birds and mammals.

Each creature is a link in a food chain of eaters and eaten. In any area, food chains connect to create a food web. All living things, plants and animals, also form a food pyramid. Green plants are at the bottom. Only green plants can make food. The smaller, middle layer consists of animals that feed on plants. The pyramid tip is made up of animals that eat other animals.

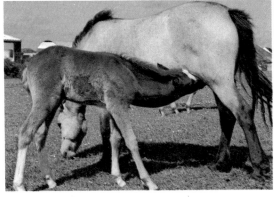

Above: A foal sucks milk from a mare. Horses are mammals. All mammals feed their young with milk from the mother. They are also warm-blooded and have hair. Man is a mammal.

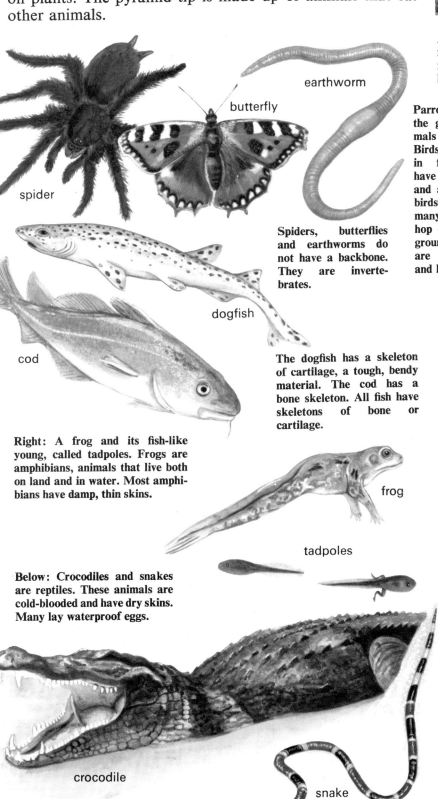

Spiders, butterflies and earthworms do not have a backbone. They are invertebrates.

Parrots belong to the group of animals called birds. Birds are covered in feathers and have claws, wings and a beak. Most birds can fly and many can walk, hop or run on the ground. All birds are warm-blooded and lay eggs.

The dogfish has a skeleton of cartilage, a tough, bendy material. The cod has a bone skeleton. All fish have skeletons of bone or cartilage.

Right: A frog and its fish-like young, called tadpoles. Frogs are amphibians, animals that live both on land and in water. Most amphibians have damp, thin skins.

Below: The koala is a marsupial. Marsupials are pouched mammals. Their babies are born so small and weak that they cannot possibly look after themselves. Instead, they spend their infancy inside a pouch on their mother's body. Most of the world's marsupials live in Australia. They include kangaroos and wallabies.

Below: Crocodiles and snakes are reptiles. These animals are cold-blooded and have dry skins. Many lay waterproof eggs.

A Lifetime of Instinct

In an experiment, this young monkey has been separated from its mother. It has been given a dummy mother made of wire and wood and covered with cloth. The animal clings by instinct to the dummy as if it were its mother.

Some animals live like machines. They act by instinct. Others can learn to do things.

All animals behave in certain ways at certain times. They perform many of these actions automatically. They do not learn how to do them. It is as though animals had been programmed like a computer. This programming is called instinct.

Some creatures live almost entirely by instinct. Insects like bees, wasps, ants and termites, for example, live in groups. They lead quite complicated lives. Yet instinct controls almost every action. Each honeybee's built-in program tells it how to look after the eggs and young inside the hive.

Instinct controls almost all actions of invertebrates, animals without a backbone. For example, they attack prey, hide from enemies and move towards or away from light, all by instinct. They cannot think about their actions because their brains are too simple.

Most vertebrates—fishes, reptiles, amphibians, birds and mammals—have better brains than invertebrates. So most vertebrates not only act by instinct, they can also learn how to do certain actions.

Even so, the early actions of a baby bird or mammal are instinctive. A tiny newborn kangaroo instinctively crawls through its mother's fur to reach the safety of her pouch. In the same way, instinct tells a newly hatched chick to crouch if something moves nearby.

In fact, birds do most things by instinct. Storks make their untidy roof-top homes and swallows build their nests of mud without actually learning how to do so. Birds also migrate, often flying thousands of kilometres, by instinct. But learning is important, too. Practice helps a young bird to improve its flying. Young birds also learn where to find food and how to recognize and avoid enemies and other kinds of danger.

Learning is especially important in the life of some mammals. For instance, practice helps lions and wolves become skillful hunters. Certain creatures—dolphins, apes and elephants—are very good indeed at learning. Trainers teach dolphins to jump through hoops. One zoo had a chimpanzee that rode a motorcycle safely through a busy city. In Burma, elephants learn how to lift heavy logs through the forests. Some even learn to muffle their bells with mud while stealing bananas.

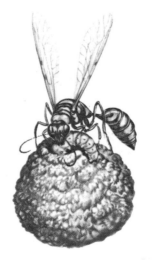

This female potter wasp has made a pot-shaped nest out of clay. Now she is stocking it with caterpillars. Next she lays an egg inside the nest, then seals the pot and flies away to die. The grub that hatches from the egg never sees its mother, so it can never learn from her how to build a nest. Yet when a young female potter wasp grows up, it makes a clay pot and stocks it with caterpillars just like its mother did. This is because the wasps work by instinct.

Above: The baby monkey clings, by instinct, to its soft cloth dummy mother while it gazes at a hard wire dummy mother. The hard wire dummy holds a feeding bottle. The baby learns to feed from the bottle. But it will only do so if it can still cling to the soft cloth dummy. The baby also learns about a heated pad that gives it warmth. But it still prefers to cuddle the soft cloth dummy. These experiments show, perhaps, that instinct matters more than learning, even to animals as intelligent as young monkeys.

Below: These pictures show how learning changes the way in which a chick behaves. In the first two pictures, the chick is very young. Instinct makes it crouch down when any large bird passes by, even if that bird is a harmless swan. In the two other pictures, the chick is older. It has learned to recognize some kinds of birds. The chick now knows that long-necked birds, like swans, are not dangerous. But it still crouches when an unknown bird with a short neck passes overhead. In fact, this short-necked bird is a falcon, a bird of prey that often attacks young chicks.

11

Keeping in Touch

Only people hold complicated conversations. But most animals can send each other simple signals. They often use sounds or gestures.

Animals talk to one another. Like people, they use sounds and gestures to pass on messages. Some also use touch and smell. Animal language can only be used for simple messages. But they are important messages: ones that help animals track down food, find a mate, escape danger and keep together as a family or group.

Above: The male fiddler crab uses its giant claw at mating time. It makes beckoning gestures to a female. Sometimes one female is surrounded by several males, each one beckoning her to come and be his mate. Rival males also fight each other with their claws.

Left: The chimpanzee in the bottom photograph is showing great attention. In the top photograph, its open mouth and bared teeth show that the animal is angry. A chimpanzee can communicate different emotions by changing the expression on its face. Man can do the same. But many animals have a mask-like face which always shows the same expression; they cannot change it. Such animals must use other ways to communicate their different moods.

The sound languages of certain animals are well-known. Everyone recognizes a dog's bark, a horse's neigh or an owl's hoot.

Mammals use sounds a great deal. An angry cat snarls and spits, but a pleased cat purrs. Cats that miaow to be fed or let out have learned a simple sound language that helps them talk to people. Usually, mammals use sounds to signal to creatures of their own kind. In the wild, hunting wolves make short, sharp barks to each other. They also howl, as a kind of get-together signal.

Perhaps the most familiar sounds in nature are the songs of birds. In most cases, it is the male bird that sings. During the spring their song can mean two things: 'This is my territory, keep out' and 'I want a mate'. At any time of year birds that fly in flocks use certain calls to keep in touch. Rooks and starlings do so. So do mother birds and their young. Birds seem to make these call sounds by instinct. Instinct also controls alarm signals like the harsh cry made by pheasants. But birds learn some sounds through listening to their parents and then copying them.

This greylag goose communicates three different emotions simply by standing in different positions. Left: The bird is about to make an attack. Centre: It is on the defence. Right: The bird's attitude shows a mixture of fear and a wish to attack.

Young finches brought up without parents, never learn the finch song. Some, brought up by other birds, learn their foster parents' song.

Many animals that live in water use sounds. Some fish seem to make noises when they are looking for a mate. Certain catfish grunt, probably to help them keep together when they are swimming in dark water. The ocean giants, whales, make a variety of noises. Scientists do not yet understand what all the noises mean. But the scientists do know that dolphins can catch fish by sending out sounds and then homing in on the echoes that bounce back from objects in the water.

Gestures and other actions are silent signals. They play an important part in the language of animals. When a dog is pleased, it wags its tail. A male green lizard hunches its shoulders to scare off rivals. Male fiddler crabs wave their giant claws to attract females. The smooth newt attracts its mate by waggling its tail. Ants 'talk' to each other by touching antennae, or feelers.

Perhaps the most complicated gesture language belongs to the honeybee. When a worker bee finds flowers full of nectar or pollen, she does a dance in front of the other workers to tell them where the food is. A round dance means that the food is near the hive. But if she dances in a figure of eight and wiggles her tail at the same time, then the food is farther off. The speed of the dance shows the distance of the food. The direction of the dance shows which way the bees must go to find the food.

Many animals pass on messages by scent. A dog, for instance, leaves its scent on trees to mark out its territory. If frightened, fish and some other animals give off chemicals that warn others of the danger. Scent plays a part in courtship, too. It helps guide male and female fish to one another. Tests show that the scent of a female moth can attract males that are many hundreds of metres away. Scents also help ants and bees to find food or to raise an alarm.

Most ants are blind, so when an ant finds food, such as a dead butterfly, it cannot show other ants where it is. Instead it leaves a scent trail (shown by arrowheads in the picture). The scent soon evaporates, but by then the other ants know the way.

Private, Keep Out!

In most cases, each animal has its territory, the patch of land or water where it lives. A lizard's territory can be as small as a few stones; a lion's territory, as big as 200 square kilometres.

Animals of the same species compete fiercely with one another for somewhere to live and feed. This is because their needs are very similar. Animals of different species compete less fiercely because their needs do not clash so strongly. Social animals—animals that live in groups—guard and defend their territories all through the year. Other animals claim territories only at breeding time.

Birds' territories are among the easiest to find, especially in spring at breeding time. In most species, each male bird claims a territory. A small bird like a robin needs no more than a garden. But a golden eagle may claim as much as 80 square kilometres (30 sq miles). Gulls and other seabirds that nest in colonies only take nesting space. Such birds do not need a large area because they find food over the wide ocean.

In spring, the male three-spined stickleback leaves his shoal and stakes out a breeding territory. His belly turns red and he swims round and round his territory, displaying his belly to scare off rivals. He builds an underwater nest where his mate lays eggs. The male guards the nest and drives off all other creatures.

Many animals stake out and defend territories. Most kinds mark their territories with scent. Dogs and foxes use urine as scent markers. Some mammals have special scent glands. Antelopes and deer, for example, mark trees with an oily scent from glands near the eyes.

Like birds, mammals try to scare away rivals. Howler monkeys make fierce booming noises to frighten off competition. If male mammals meet at the edge of each other's territory, each wants partly to fight and partly to run away. Instead of doing either, they may begin to feed, or wash, or gather nest material. Even if they do fight, it is rarely violent. In fact, most territorial fights are bluff and end when the weaker animal retreats, unharmed.

Some territorial fights, however, are very fierce. Like the fights between male sea elephants. At breeding time each bull, or male, claims a small area of beach where it collects

a group of cows. Rival bulls then fight fiercely. Some are badly wounded. A few even die.

Fish normally defend territories during the breeding season. They cannot stake their claims with scent or song, but they can and do display in special ways. Siamese fighting fish make themselves look hideous by holding all their fins erect. Cichlids open out their gill flaps in a menacing manner. At all times the brightly coloured bodies of many fishes found on coral reefs act as warning signals to would-be trespassers.

Among the reptiles, crocodiles and lizards are both territorial. A female crocodile defends the nest of rotting vegetation in which her eggs are laid. A male lizard claims a clump of bushes and chases off rival males.

Some social insects also attack rivals. An ant that wanders into a nearby ants' nest may be quickly killed and eaten by its unfriendly neighbours.

The territory claimed by a bird depends largely on the bird's size and on the amount of ground it covers in its search for food. In one area a different species, with different food needs, can guard overlapping territories. The territory of a Baltimore oriole (shown in yellow) is quite small because orioles do not have to fly far to find insect food. A barn owl eats small mammals such as mice and voles and must therefore cover a far wider area (red). Of the three birds shown here, the bald eagle needs the largest territory (blue). This huge bird hunts fish as well as lambs, kids and fawns. Its territory has to be enormous to provide enough food for itself and its family.

By rubbing its head against a branch, this male antelope smears the branch with scent from a gland near the eye. The scent warns other males that they have reached a rival's territory. They often retreat rather than risk a fight.

The Mating Game

A Prince Rudolph's blue bird of paradise. The magnificent male woos the rather dull-coloured female by showing off his brilliant plumage and then by dancing and strutting about.

Generally, a male animal and a female animal must meet and mate before the female can lay fertile eggs or give birth to babies. If the males and females of an animal species never mate, then that species dies out.

Once an animal has found enough food, its next need is to find a mate. This is no problem for animals living in flocks or herds. Bison and wildebeeste, for instance, can always find a mate nearby. But for creatures that live solitary lives, finding another member of the same species may not be so easy.

Solitary animals often use special signals to attract a mate. Such signals may be sounds, smells or visual displays. Male crickets and grasshoppers attract females by making chirping sounds. Female glow-worms and fireflies attract males by glowing brightly at night. Some female fish attract males by giving off a special chemical from their bodies into the water. In the same way, female moths give off a chemical that is carried by the wind to distant males. Usually people cannot smell this scent at all. But male moths can follow it for more than one kilometre, even when the air is full of other strong smells.

Once a mate is found, she must be wooed. Many male animals court their females by displaying their bravery and handsome features. Some give their chosen females gifts. A male penguin, for example, rolls a stone towards the female he is wooing. Jackdaws offer food or nest material. Some males attract their females with a home. A male heron begins building a nest before he looks for a female. The bower bird makes a remarkable tent decorated with bits of glass and berries.

But birds court mainly by display. A male bird struts and shows off his gay plumage to the hen which is usually a duller colour. Some male birds perform ritual dances. Each kind of duck has its own special dance. This makes sure that females choose males of their own kind and not ones of another species. Some birds, like great-crested grebes, dance together. In some cases, male birds attract females by singing.

Courtship is less showy among mammals. Stags attract female deer by loud barking noises. But they spend less energy courting females than fighting off rival males.

Among reptiles, male tortoises woo females by biting their head and front legs. Indeed, with many animals a male's courting behaviour is very aggressive. Females sometimes overcome this aggression by making a display of appeasement. Such displays include ritual dances and offering food. Acts like this help to reduce the male's aggression and remove the female's fear. Only when both these things have happened can the creatures mate.

Many animals pair off just for the breeding season. A few mate for life. Two animals who have paired for life, frequently perform the ritual acts that helped to form the bond between them. In this way, they strengthen and keep alive their pair bond.

Above: The female emperor moth uses scent to find herself a mate. She produces the scent from a special gland. The scent travels with the wind and is picked up by a male's feathery feelers or antennae. The male follows the scent by flying upwind and so manages to find the female.

Right: A Siamese fighting fish swims below his bubble nest and displays his long exotic fins to attract a female. Then he wraps himself round her and they mate. After mating he blows her eggs into the nest and guards them.

Left: This male tree frog attracts a female by uttering a special cry. He fills his throat with air, then forces out the air to make a high-pitched call. Female tree frogs find this sound irresistible.

Below: Two male antelopes clash horns in a fight at mating time. The fight looks fierce, but such battles usually end with the weaker animal running away unhurt. The winner mates with the females in the loser's herd.

Building with Mud, Wax and Paper

swallow

Many animals build a shelter for their young. Some of the most fascinating are made by birds and insects.

Flamingoes build mud nests that stand in shallow water. The top of each nest is shaped like the inside of a saucer. The mud soon hardens. Then each female lays one egg inside her nest. A few days after hatching, the chick jumps down into the water.

The swallow builds a cup-shaped nest on a ledge in a building. To make the nest, the bird collects mud in her beak and mixes it with grass. She lines the nest with feathers to make it soft and warm.

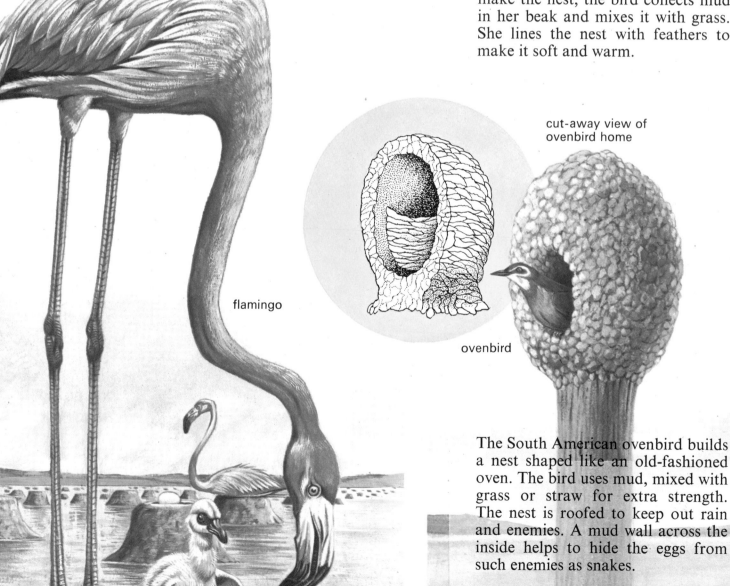

flamingo

cut-away view of ovenbird home

ovenbird

The South American ovenbird builds a nest shaped like an old-fashioned oven. The bird uses mud, mixed with grass or straw for extra strength. The nest is roofed to keep out rain and enemies. A mud wall across the inside helps to hide the eggs from such enemies as snakes.

potter wasp

The mud dauber builds a nest from balls of mud. In each mud cell she lays one egg. Then she places paralyzed spiders inside the nest with her eggs. When the eggs hatch, the young grubs feed off the spiders.

The female potter wasp constructs a tiny mud pot on a twig. She lays one egg inside the pot. Then she paralyzes a caterpillar with her sting and pushes it into the nest. In this way the wasp provides fresh food for her baby grub.

mud dauber

paper wasp

close-up of wasp's head

mandibles for chewing

A paper wasp softens bits of wood by chewing them. Then it builds a papery nest hanging from a branch.

A worker honeybee produces wax in its body. The wax comes out of tiny holes in its abdomen. The bee uses its jaws to shape the wax into six-sided cells that form a honey-comb.

honeybees

19

Outwitting the Enemy

Some animals are hunters. They often have sharp teeth or claws for killing their victims. The victims, other animals, have different ways of defending themselves.

A kangaroo rat escapes from a rattlesnake. The snake strikes but the kangaroo rat leaps into the air, using its long back legs as springs. This high jump often saves the rodent's life.

Some animals have built-in weapons to defend them from enemy attacks. For example, a cow has horns and a tortoise has its shell. Other animals have no such powerful defences. Instead, many of them survive by simply running away. The urge to run away from attack and danger is an instinct. When hounds chase a fox, the animal automatically flees. By instinct, the escaping fox tries to cross water as this breaks the scent trail it leaves behind. But foxes also learn cunning. A fox that has been chased before may double back on its own tracks. The animal knows that this tactic confuses its enemies.

Large plant-eating creatures, such as deer, antelopes and horses, are less crafty than the fox, but they run fast enough to escape most enemies provided they have sufficient warning. As these animals feed, they constantly sniff the air and gaze round, on the watch for danger. A strange scent can send a herd galloping away in fright. This is why hunters try to creep up on deer from downwind: with the wind blowing their scent away they can take the animals by surprise.

Because they are so low on the ground, rabbits, hares and marmots cannot see as far as horses. These small mammals sometimes rear up on their hindquarters to look and listen for danger. A hare that sees a man crouches down and often escapes unnoticed. If the man comes very close, the hare races away almost as fast as a car.

Some animals escape their enemies by leaping, climbing or burrowing. If a hunting creature pounces on a jerboa or a kangaroo rat, the small animal leaps into the air so that its attacker misses. This action often gives the hunted animal just enough time to bound away. Squirrels and cats escape their enemies by climbing trees. But when a cat is cornered and cannot run away, it turns and fights with teeth and claws. Badgers, foxes, moles and rabbits escape their enemies by diving underground down a burrow. Few animals can follow them.

With their long horns, gemsboks look fierce. But when an enemy comes near, these big antelopes prefer to run away instead of fight. They can gallop faster than most of their attackers.

A flock of starlings flies close together when a bird of prey, like a falcon, approaches. Such a thick mass of moving birds makes a difficult target.

Many animals rely on camouflage to hide them. Fish tend to have dark backs and silvery bellies. Seen from above, their backs blend with the water. Seen from below, their bellies blend with the sky. A threatened flatfish hides its body with a covering of sand.

Some creatures have ways of distracting an enemy. A frightened squid squirts out a cloud of ink. This hangs in the water and confuses the enemy while the squid escapes. Many lizards shed their tails if attacked. While the attacker grabs at the twitching tail, its owner scuttles off unharmed.

The chuckwalla has an unusual defence. This lizard hides among rocks. It blows up its body until it totally fills a rock crevice. Then, no enemy can pull the lizard out.

Pretending to be dead is another defence. The grass snake, the opossum and some spiders play dead when danger threatens. Enemies move off, leaving the prey unharmed.

Left: This lizard shed its tail when the bird of prey attacked. While the bird is distracted by the wriggling tail, the lizard escapes. Later, the lizard grows a new tail, but this tail is shorter than the first one and a different colour. Sometimes lizards shed their tails when they are fighting each other.

Right: A startled cuttlefish squirts out ink from a bag inside its body. As the cuttlefish swims away through the water, the cloud of ink remains just long enough to fool an enemy into thinking that the ink is the animal itself. The cuttlefish's relations, the squid and octopus, also shoot out ink in this way.

21

The Long Winter Sleep

In winter, many animals cannot find enough food to fuel their bodies. To save energy, some of them sink into a long, death-like sleep during the cold months. This sleep is called hibernation.

People need regular hours of sleep. So do most mammals and birds. When creatures fall asleep, the workings of their bodies slow down so that they are just ticking over like a car engine when it is idling. Not all animals sleep in the same way. Some of them sleep lying down, some standing, some sitting. Certain animals need more sleep than others. An elephant manages with only half an hour's sleep actually lying down.

Most creatures sleep once or twice every twenty-four hours. Some also have a long winter sleep called hibernation. Where food is scarce in winter, these animals hibernate for several months, or even for half the year.

Hibernation is not like ordinary sleep. When a mammal hibernates its temperature drops very low. At the same time its breathing and heart beats slow down. The creature becomes so still and cold that it seems to be dead. In this state its body uses very little energy. So it gets enough energy from food stored as fat inside its body. In autumn, animals that are going to hibernate eat so much that they become round and plump. Dormice, hedgehogs and woodchucks, for example, all grow plump. In winter they hide and live off their body fat. Next spring, after the long sleep, they are thin and hungry.

The fourteen animals shown on these pages all hibernate or at least, sleep, in winter. In the picture, the woodchuck and the dormouse are the only two mammals that truly hibernate. The bear and squirrel, for instance, come out on mild days to find food. Cold-blooded creatures, like lizards, snakes, tortoises, newts, salamanders and toads, only hibernate in cold countries. In hot countries, there is no need for them to hibernate. In the insect world, most adults die during the winter and only their eggs or pupae survive. But there are some butterflies that manage to live from one year to another. Earthworms usually burrow deep down in the ground to escape the cold. When they curl up for their winter sleep, they do so in moist soil to stop their bodies becoming too dry.

newt

natterjack toad

adders

green toad

woodchuck

common toad

tortoise

Hibernating animals spend the winter in a dark, quiet place. Even butterflies, insects that love the sun, hide in dark corners. Some, like the peacock and tortoiseshell, have drab undersides to camouflage them.

Many hibernators hide underground where they are protected from snow, ice and frost. Lizards, snakes, toads and tortoises all burrow in the soil before its surface freezes. Fish that hibernate dig themselves into the river bed. In California, one kind of nightjar spends the winter sleeping under boulders. Most birds, however, survive cold weather by finding enough food to keep their bodies active, or by flying off to warmer lands.

Some animals that sleep a lot in winter do not truly hibernate. On mild winter days sleeping bears wake up and leave their dens in search of food. In the very coldest weather even hibernating animals wake up and move about so as to raise their body temperature. Otherwise they would grow cold enough to freeze.

brown bear

Camberwell beauty butterfly

dormouse

red squirrel

salamander

earthworm

sand lizard

A horseshoe bat hibernating in a dark cave.

The African lungfish burrows into a muddy river bed as the water dries up. The fish sleeps in the sun-baked mud until the river flows again. Then it swims out. Many creatures sleep through hot, dry weather. Frogs, toads and some small rodents that live in deserts survive months of summer drought and heat by sleeping. This kind of summer sleep is called aestivation.

23

Animal Playtime

Animals spend their lives finding food, escaping danger and sleeping. Yet some mammals also find time for games.

Play teaches children how to behave with other children. And it helps them discover how things work in the world around them. Play also helps some other mammals to learn. Many of the games played by young animals appear to be practice for a grown-up activity. Lambs frisk and jump as if learning to climb a rocky mountain. Lion cubs pretend

Left: A brown bear cub playfully wrestles with its mother. Mock fighting strengthens the cub's muscles. It also helps to develop the skills that the young bear will need later for hunting.

to fight each other. Bear cubs wrestle. Such activities are good practice for fighting off enemies in later life. A puppy leaps on a rug, grasps it in his jaws and shakes it as if it were a rat. A kitten creeps up on a dead leaf or ball of wool and then pounces on it, just as an adult cat stalks and kills a mouse.

Some young mammals seem to learn to hunt, to fight and to escape from enemies mainly by copying the actions of their parents. Many parents teach them these actions in play. A mother cat, for instance, shows her kittens how to pounce on prey and kill it.

Not all young animals play. They certainly do not play to learn. Instinct, not learning, controls the way in which they behave. Indeed, tests show that not even cats have to learn to chase and kill their prey. This happens through instinct. It seems that young creatures play mostly to use up spare energy.

Adult animals generally use up so much energy finding food and escaping from enemies that they have little energy left for play. Yet the adults of some species seem to love a game. Otters enjoy sliding down a bank of mud or snow into the water. They also toss pebbles into the water and dive after them. Badgers, too, are playful beasts. People have seen them playing leapfrog, king-of-the-castle and even using balls of mud as toys. Perhaps adult dolphins are the most playful of all. These water creatures perform acrobatic leaps for fun. Tame dolphins appear to enjoy jumping through hoops and walking across the water on their tails.

A badger family at play. Badgers often use stones and lumps of mud as balls. In the picture, one of the cubs has discovered a much more fascinating plaything, a hedgehog that has curled up for protection. Two other cubs are having a pretend battle. Later on, the adult badger on the right may join its young for such games as tag and leapfrog. Badgers and their relatives, the otters, are among the most playful of all mammals.

The twitching of its mother's tail awakens this young cheetah's hunting instinct. The cub's playful attacks on the tail improve its hunting skills. But basically these skills come by instinct. For the young cheetah the main purpose of play is to use up spare energy. Adult cheetahs are less playful because they have less energy to spare; they use up a lot of energy in their search for food.

Adult animals in zoos seem to spend more time at play than adults in the wild. As they do not need to hunt for food they have more time and energy for playing. Play helps the animals to keep their bodies fit and use up the energy they would have spent in hunting. Chimpanzees in captivity invent games in which they roll, slide and somersault. Because they are intelligent, chimpanzees enjoy more complicated games than most other animals.

Compared to many other animals, birds do not seem playful. Crows and parrots are the main exceptions. A captive parrot can lure a person to its cage then peck him hard. Jackdaws and ravens enjoy riding a strong wind, letting it sweep them up into the sky, then closing their wings and diving down. Even sparrows seem to have a sense of fun. In spring they jump on daffodils, until these sag. They hop off to let the flowers spring up. Then hop on for another ride.

Animal Spotting

Animals are everywhere. Whether at home, in a park, by the sea or in the country, people can watch animals and learn about the way they live.

In everyday life, there are many opportunities for watching and studying animals. There are pets at home. There are birds in parks and gardens. And, in the country, there are many different creatures living in woods and fields and hedges.

Keeping pets can be very worthwhile. As well as cats and dogs, there are other animals such as mice, gerbils, guinea pigs, hamsters and canaries. But having a pet, even a goldfish, is a responsibility. Pets have to be fed, cleaned and generally looked after.

A good way of watching birds is to hang up nuts or fat outside the window. Strings of nuts are very popular with tits. They cling onto them, often upside down, and peck away. Other birds eat bits of bread, fruit and seeds. This type of food can be scattered on a bird table. Birds love to bathe in a bird bath. They splash happily in the water,

then shake and clean their feathers.

A country walk in spring is a good time for finding out about nests: where birds nest, how they build their homes, how they feed their young and how they catch their food. Some catch insects in the air. Some hunt for worms in the grass or in plowed fields. Some feed on certain fruits and seeds.

Wild mammals are more difficult to watch than wild birds. Many only move about at night. Others live in long grass or fallen leaves where they are difficult to see. Anyone watching wild animals has to have a great deal of patience. It is often necessary to sit still and quiet for a long time, preferably wearing dark clothes and hidden by a bush or fence. At night, it is possible to watch wild animals with a flashlight, but the glass must be covered by red cellophane or plastic.

cardinal bird

badger

frog

duck

mole

	cardinal bird
	owl
	badger
	deer
	frog
	duck
	mole
	pheasant
	fox

owl

deer

pheasant

fox

Most wild animals are so timid that people seldom see them. But one way of learning about their habits is to study the tracks they leave. Animal footprints show up quite clearly in soft mud, sand and snow. The depth and position of each track is a clue as to what the creature was doing. It may have been walking or running. Perhaps it was chasing prey or escaping from an enemy. Maybe it was on its way to its nest or burrow.

The best way of identifying tracks is to make a drawing of them or, better still, a plaster cast and then compare the cast or drawing with pictures of tracks in a nature book. It is also important to keep a record of where and when the tracks were seen.

Tracks are not the only clues animals leave behind. There is always plenty of evidence around of their eating habits. Chewed grass, stems or nutshells show where a plant-eater has been feeding. Bones and bits of fur or feathers show where a flesh-eating creature has had a meal.

Animals leave many clues behind them. Deer, foxes, ducks and many other creatures all have different footprints. Gnawed bark on trees shows where deer or rabbits feed in winter. Nuts split in half may have been broken by a squirrel. A nutshell with a neat hole in it has probably been chewed by a mouse. Owls spit out pellets containing bits of bone and fur and beetle's wings.

The smaller members of the animal world are also worth watching. Tiny creatures like insects, slugs and spiders lead fascinating lives. It is quite easy to find a spider and study it as it spins a web and then traps flies in it.

Looking is just part of finding out about animals. Listening is another part. Birds and mammals make various sounds. Many of these calls, cries and songs have been recorded. Hearing them on a record or cassette is a good way of learning to recognize the voices of foxes, squirrels, shrews and many other creatures.

A very important aspect of finding out about animals is making notes. These notes should cover when and where each creature was seen and what it was doing. This type of record helps give a complete picture of how and why wild animals behave in the ways they do.

27

In the Beginning

The first plants were simple water plants called algae, from a Latin word for seaweed. Algae have been in the oceans for over 3000 million years.

Some of the world's first plants were algae. Living algae include seaweeds and the tiny plants that turn pondwater green and choke ditches.

Like most other plants, algae use sunlight to help them make food, mainly from chemicals in the water. Because they live in water, algae need less complicated bodies than most land plants. A seaweed, for instance, is not divided into proper roots, stems and leaves. Inside its body, there are no tubes to carry food and moisture.

Some seaweeds have no special shape. Sea lettuce is one example. On the other hand, oarweed has a long, waving 'leaf' that looks like a strap. In fact, many seaweeds do have parts that resemble true leaves and stems. There are some seaweeds with feathery leaves. Others have branchlets like the teeth of a comb. Then there are seaweeds with stiff, brittle stems. Bladderwrack has flat leaves with pea-sized air bladders. These bladders serve as floats to lift the seaweed when the tide comes in.

Seaweeds have no roots, but they do grow rootlike holdfasts that help them grip the sea bed. Some seaweed holdfasts are shaped like claws. Some are round. Some are

Above: Some common seaweeds. Sea lettuce (1) and *Enteromorpha* (2) are green seaweeds. Channelled wrack (3), bladder wrack (4) and toothed wrack (5) are brown seaweeds. So are *Pylaiella* (7) and the two big laminarians (10) and (11). The red seaweeds are *Porphyra* (6), *Dilsea* (8) and *Heterosiphonia* (9). Below: This Japanese woman sorts seaweed gathered on the sea bed. Seaweed is rich in nourishment and the Japanese eat a great deal of some kinds.

sticky threads. Holdfasts help seaweeds cling to under-water rocks in the fiercest storms.

All seaweeds need the energy in sunlight. But sunlight is made up of all the colours in the rainbow. Green sea-weeds need the red in sunlight. As red rays only penetrate the surface, green seaweeds cannot grow deep down. Red seaweeds, however, use blue light. This penetrates water deeply so red seaweeds can live lower down than others. Brown seaweeds are found at middle depths. All seaweeds are green or red or brown.

All seaweeds have small beginnings. Some start as plant-lets growing on their parents' 'leaves'. Others sprout from runners thrown off by the parent plants. And others begin as spores produced in special branches on adult plants. Many seaweeds remain small, but some become very big: one brown seaweed grows over 100 metres (330 ft).

Seaweeds live in water but some algae live on land. Little *Protococcus* grows on fences, tree bark and damp soil. Thousands of these tiny plants make up a green film. This alga thrives best in moist, shady places.

Scientists divide algae into two main groups: blue-green algae and all other algae. Blue-green algae belong to the animal kingdom; all other kinds of algae belong to the plant kingdom.

Blue-green algae are tiny beings made up of just one cell. Like all other algae, the blue-greens contain the green substance, chlorophyll, that helps green plants make food. But unlike plants, the blue-greens usually multiply by simply splitting in two. Also, their cells do not have the special structure called a nucleus. These facts show they are related to another group of tiny, one-celled organisms, the bacteria. Most bacteria feed off substances that are, or were, alive. Bacteria were probably the first living things on Earth.

Mushrooms and Molds

Fungi are peculiar plants. They have no roots, stems or leaves and cannot make food. Many of them live in the dark.

Fungi include mushrooms, toad-stools, yeasts, molds and mildews. Fungi differ from ordinary plants in many ways. They have no roots, stems or leaves. In fact, they have no true plant body. Instead, they usually grow as a mass of thin threads. In some fungi these threads form long tubes. In other fungi the threads are divided into separate cells.

Fungi also have no chlorophyll. This is the substance that makes green plants look green and helps them to make food using the energy obtained from sunlight. Fungi cannot make food in this way. They have to get food ready-made from green plants and animals. So fungi do not need light to grow. Indeed, many live in darkness. For these reasons botanists place fungi in a group of their own instead of putting them with ordinary plants.

Fungi feed in various ways. Some feed on dead plants or animals or on animal droppings. These fungi are called saprophytes. Others feed on living plants or animals. These fungi are called parasites, and the plants or animals they feed on are known as their hosts.

Fungi have special ways of feeding and growing. They produce substances that help them to turn their food into liquid they can digest. The fungi's threads take up food and grow. They lengthen and branch out so that they spread through and over the food supply. This mass of threads is often known as spawn. Botanists call it mycelium. If a bit of mycelium is broken off, it may become a separate fungus.

New fungi usually begin life as spores. A spore is just a single cell. It lacks the food supply found in seeds produced by flowering plants. Insects, or the wind, carry away the spores of fungi that grow on land. Water fungi produce spores with tails that help them swim away. When a spore reaches a food supply it starts to divide and grows into a new fungus. In fact, what actually happens is quite complicated. Fungi usually grow two kinds of thread.

Above: The mold *Aspergillus* greatly enlarged. One head is cut through to show how the spores grow. The mold grows on leather, food and other materials.

Above: the earthstar fungus grows in beech woods. The outer layer splits and spores are then released from the centre core.

From left to right, the pictures show how the death cap's fruiting body develops. The fruiting body starts to grow inside a skin. In time, the stem and cap burst out. The skin remains as a cup round the bottom of the stem and another skin forms a ring below the gills. When the gills ripen, they release millions of spores. The skin cup and ring on the stem, plus the white colour of the gills, help distinguish the death cap and other fungi in its group. They are all poisonous and must never be eaten.

Bracket fungi growing from a tree trunk. The fungus draws food from the sap of the heart-wood. Trees attacked by bracket fungi often die. The tree in the picture is already showing signs of damage.

Fly agaric, also called fly *Amanita*, is a very colourful fungus. Its cap is red, yellow or orange. This fungus grows near conifers and birches. Fly agarics feed on tree roots and decaying substances. Although the fungi look good enough to eat they contain a deadly poison. This poison paralyzes the nerves that make the heart work.

Only when both kinds meet and join can they produce a so-called fruiting body. This is where the spores are formed. So, in a way, fungi reproduce like flowering plants where male and female cells join to make a fertile seed. Wheat rust has a particularly complicated way of breeding. This fungus produces five kinds of spore at various stages of its life. At one stage, it produces spores on wheat plants; at a later stage, it produces spores on barberry plants.

Mushrooms and toadstools are the best-known fungi. The part above ground is the fruiting body that produces spores. The mycelium is underground. Most fungi feed on manure or rotting plants in the soil. But some eat living wood. Bracket fungi can kill a tree.

Rust fungi, many mildews and some other fungi attack crops and animals. Athlete's foot, dry rot, potato blight and Dutch elm disease are all caused by fungi.

Molds grow on food, paper, leather and other materials produced from plants or animals. Certain molds produce the useful group of drugs called penicillins. Others create blue, black and white growths on stale bread.

Yeasts are one-celled fungi that can quickly multiply and turn sugar into carbon dioxide gas and alcohol. In bread-making, yeast produces the carbon dioxide that makes dough rise. In wine-making, yeast produces alcohol.

LICHENS

A lichen is really an alga and a fungus: two plants in one. The alga is a green plant that uses sunlight to make food for the entire plant. The fungus is able to take in water and chemicals, again for the whole plant, from the earth or rock. The fungus also protects the alga. In this way, lichens can live in conditions that are too harsh for just an alga or a fungus or any other plant. Lichens are plant pioneers. When they grow on bare rock they start to break it up. By doing this, they gradually form soil where other plants can grow.

Feather Ferns and Velvet Mosses

The first true land plants were related to ferns and mosses. They began growing on the damp edges of lakes and rivers 400 million years ago. Today, most ferns and mosses still need to grow on moist soil or stone.

Early plants lived in water. They had no roots, or stems strong enough to hold up leaves in air. They had no true leaves and no tubes to carry food and water through the plant. Large land plants need these things.

Pteridophytes were the first land plants with well-developed stems and leaves, true roots and tubes to carry food and water. Pteridophyte means feather plant. Pteridophytes include ferns, plants with feathery leaves. Club mosses and horsetails are also pteridophytes. In ancient times there were forests of giant horsetails, ferns and club mosses. Yet today, most ferns, club mosses and horsetails are small enough to step upon or brush aside.

Mosses and liverworts belong to another group of early land plants. This group is called the bryophytes. Bryophytes have simple stems and leaves but no true roots. The plants are anchored to the ground by hairy roots called rhizoids. Bryophyte roots, stems and leaves have no tubes to carry food and water from one part of the plant to another. This helps explain why liverworts and mosses are never very large.

A horsetail, a relation of the ferns. Fertile stems (above) have cones where spores develop. The feathery green stems (below) have no cones.

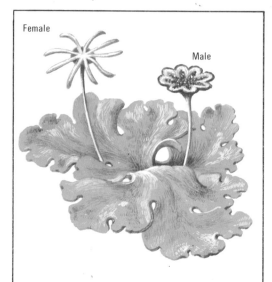

LIVERWORTS

Liverworts are small, low-growing plants found only in moist places. These pictures show examples of the two main kinds.

Above: A liverwort with a leathery body that lies more or less flat on the ground. The upright structures produce male and female cells that play a part in making new liverwort plants. New plants also grow as buds sprouting from parent plants.

Below: A leafy liverwort. Leafy liverworts have rows of leaves growing from the stem.

Left: The common clubmoss is a small, lowgrowing relative of the ferns.

Right: Two kinds of moss. The left-hand moss is one of a group of mosses found in wood, bogs, gardens and among rocks. The other is a woodland moss. The pods are spore capsules.

The maidenhair fern grows on damp rock and in shady woods. Its leaves are in two parts. This fern gets its name from its slim, shiny stalks.

A giant tree fern from the island of Tahiti. Tree ferns grow mainly in the tropics. Unlike most ferns, their stems do not grow underground.

Mosses usually grow like a carpet on damp, shady ground. But some live on walls, roofs and rocks. Mosses are plant pioneers. Sphagnum mosses form floating carpets on lakes. The lower parts of the plants die but the tops keep growing. In time, moss fills the lake and changes it to a bog. Finally, the weight of moss above squashes the moss below into peat. People burn peat as fuel and use it to improve garden soil.

Mosses have a rather complicated life. Each plant has a leafy stem with a leafy tip. Some tips contain male cells called sperms. Others hold a female egg cell. When this ripens it gives off a sugary substance that attracts the sperms. If the surface of the moss is wet they swim to the egg cell. A sperm fertilizes the egg cell, which starts to grow. But the egg cell does not make a new moss plant. It produces a capsule full of spores. When these ripen they fall out. If they land on damp ground, they germinate and grow into new plants.

Liverworts and ferns also reproduce in two steps: one step involves male and female cells, the other step involves spores. In mosses, the main plant is the one producing male or female cells. In ferns, the main plant is the one producing spores. A fern spore grows into a tiny plant with male and female cells that produce a main plant. There are about 10,000 kinds of fern in the world.

A prickly shield fern with three feathery leaves and a new one unfurling. This fern grows in woods and on mountains in the northern hemisphere.

A male fern with some young and some full-grown fronds or leaves. Next to it is an enlarged picture of the tiny plant that has developed from a spore.

Cones and Needles

Seed-bearing plants first grew on Earth 350 million years ago. They were the ancestors of cone-bearing plants, or gymnosperms. These are all woody plants, mostly trees.

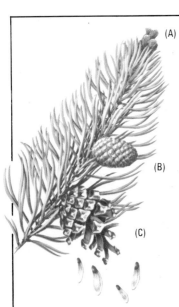

The crown, or stem top, of a male cycad. Young leaves grow upwards in the centre; cones, containing pollen, hang from the edges. Cycads are among the oldest living gymnosperms or cone-bearing plants. In the far-off days when dinosaurs roamed the Earth, cycads may have given rise to flowering plants. A cycad cone and a large flower both seem to be made in a similar way.

Most of the plants people see around them are green plants. Green plants make their own food. They make it from chemicals in air and water, using the energy in sunlight. Most green plants grow from seeds. In seed plants, tiny male cells in pollen grains join female cells in ovules. This produces fertile seeds that grow into new plants.

There are two great groups of seed-bearing plants. These are the flowering plants and the cone-bearing plants, or gymnosperms. Gymnosperm means naked seed. Unlike flowering plants, the cone-bearing plants do not grow seeds surrounded by a fruit. Most gymnosperms produce pollen and seeds in cones instead of in flowers. Each cone has either male or female cells. Often both sorts of cone grow on one plant. But male and female cones grow on separate plants in cycads, gnetales and the ginkgo. In gymnosperms, the wind carries the pollen grains to the ovules. Later, the wind scatters the fertile seeds. Wings on the seeds help to carry them far and wide.

Cone-bearing trees, called conifers, form the main group of gymnosperms. There are over 500 kinds of conifer. They include cedars, cypresses, firs, junipers, pines, spruces and redwoods, the largest trees on Earth. Most conifers are evergreen: they shed old leaves a few at a time and do not drop all leaves in winter. Conifers have small needle-like leaves which lose less water than ordinary leaves. This helps them to grow in lands too cold or too dry for other trees. ·The great forests of Canada, north Europe and Russia are all coniferous.

PINE CONES

(A)

(B)

(C)

In pines, male cones are small and soft. Female cones sprout on the tips of shoots. Young female cones (A) grow upright with open scales to let pollen reach the ovules. After pollination, the scales harden and close and the cone stalk bends (B). Safe inside the closed cone, pollen cells fertilize the female cells and the ovules develop into seeds. When these are ripe the cone scales shrink and open. This frees the winged seeds (C). In most kinds of gymnosperm, or cone-bearer, the female cones form and ripen in one year. But in pines this takes two or three years to happen. Some trees produce short cones but certain species grow cones as long as a man's arm.

Above: *Welwitschia* grows only two leaves but they are torn and split by wind and sand. This desert plant lives for over 100 years.

Right: A giant redwood tree. Some grow 100 metres (330 ft) high and up to 10 metres (33 ft) thick. Redwoods can live for 4000 years.

The oldest living gymnosperms are cycads. Cycads have existed for over 200 million years. They are tropical plants, rather like palm trees, with huge fernlike leaves. Each plant has one or more cones at the top of the stem. The cone can be enormous: up to a metre long (39 in) and weighing 45 kg (100 lb).

The ginkgo, or maidenhead tree, is another ancient plant. Fossil remains show that it has changed little in the last 150 million years. The ginkgo comes from China where it is considered sacred. It has small, fan-shaped leaves, yellow flowers and fleshy cones rather like large berries. Yews and junipers, both conifers, also produce berry-type cones.

Although gnetales are gymnosperms, they are quite similar to flowering plants. Their male parts look more like simple flowers than cones and their female ovules are guarded by leafy scales. Some gnetales are climbing plants found in tropical rain forests. Some are wiry shrubs belonging to dry places. Welwitschia is a strange desert plant from south-west Africa.

Seven cone-bearing trees. From left to right: Spruce, monkey puzzle, cedar of Lebanon, Scots pine, ginkgo, yew and juniper. Four-fifths of the world's timber comes from cone-bearers.

Buttercups to Beech Trees

Flowering plants are called angiosperms, from Greek words meaning seed container. Seeds are protected in a seed coat, or ovary. A ripe ovary is a fruit.

Flowering plants, or angiosperms, were the last main plant group to develop. They first appeared about 150 million years ago. This is not long compared with the total time plants have been on Earth. (The first plants grew over 3000 million years ago.) Today, there are some 250,000 kinds of flowering plant—more than any other kind of plant. They range in size from giant trees to tiny weeds.

Flowering plants have spread so successfully because of the way in which they reproduce themselves. Many flowering plants have brightly coloured petals. These attract insects that feed on nectar and pollen. Some pollen sticks to the insects' bodies. As they fly around, the insects spread the pollen from one flower to another. In this way the seeds are fertilized. But not all flowering plants are fertilized by insects. Some have green flowers that do not attract insects. In these cases, the pollen is carried to other plants by the wind.

Whether they have green or showy petals, all flowering

These seven objects represent seven dicotyledon flowering plants. The tree is a beech. From left to right the other objects are the flower of a rose, the fruit of a strawberry, the swollen root of a carrot, buttercup flowers, the fruit of a cucumber and the flowering head of a lupin. Altogether there are more than 200,000 dicotyledon species.

plants produce seeds in the same way. Their seeds develop in the safety of some kind of fruit. This is what makes flowering plants different from all other plants.

There are two main groups of flowering plants. They are the monocotyledons and the dicotyledons. Monocots have only one seed leaf. Dicots have two seed leaves. There are also other differences. Monocots have long, narrow leaves with veins growing side by side. Such leaves have smooth edges and a simple shape and they usually grow from the base. This is why grass keeps growing after it is mown or grazed. Most dicot leaves are broad with complicated, branching veins. The leaves may be in one piece or several pieces. A leaf with several pieces is called a compound leaf. It is made up of leaflets or groups of leaflets. Dicot leaves grow from the tip.

Dicot and monocot stems also differ. Both have tubes and fibres that carry food and water in the plant. In a dicot stem, bundles or tubes and fibres form a ring round the middle. During the growing season, woody dicots add a new ring of bundles. In a monocot stem, the bundles of tubes and fibres are scattered. In each growing season, a monocot stem becomes longer, but not thicker. Monocot petals are usually in threes. Dicot petals are in fours or fives.

Monocots include bananas, palms, lilies and grasses, both wild and cultivated. Dicots include broadleaved hardwood trees and most shrubs and herbaceous plants. More than four out of five flowering plants are dicots.

The eight objects on this page represent eight kinds of monocotyledon flowering plant. From left to right the lower illustrations show an ear of wheat, a swollen onion stem, the fruit of a banana, flowers of a daffodil and a lily and the fruiting stalk of a pineapple. Above this is a sisal plant. Beside that is a palm tree.

What's In a Wood

Trees are the main plants found in woodlands. But many others grow around them.

Woods grow where the soil is moist enough, deep enough, and warm enough for trees to put down roots. Trees are easily the largest woodland plants. Outside the tropics there are two main types of woodland tree. Most trees in the world's cool northern forests are conifers such as firs, larches, pines and spruces. Apart from larches, most conifers are evergreen. These trees can survive long cold winters and short summers. In winter, snow slips easily off their needle-like leaves. The leaves' thick skins protect them from frost and from drying up. When spring comes, the

red squirrel

Right: Coniferous woodland. Few plants grow well beneath conifers, but their seeds provide food for animals such as red squirrels. Far right: A deciduous woodland with a rich layer of low-growing plants. Grey squirrels live here.

beech

common oak

ash

elm

Above: Eight kinds of tree, each shown with its own leaf and seed case. The beech, oak, ash, elm and silver birch are broad-leaved trees; they shed all their leaves in the autumn. The larch also sheds its leaves but it is a conifer. The Scots pine and the Norway spruce are conifers that stay evergreen. In both types of woodland, one or two tree species are usually more plentiful than the rest. So people talk about ash woods, beech woods and oak woods. The chief or dominant kind of tree affects all other plants growing in the wood. This is mainly because different trees cast different amounts of shade. Few plants grow in the shady soil under beech trees. But there are several plant layers under oak and ash trees.

Left: Cushions of moss belong to the lowest layer of plants found in woodlands.

Right: Eight herbaceous plants from the field layer in woodlands of broadleaved trees.

Below: Hard ferns growing in a wood. Ferns also form part of the field layer under trees.

wood sorrel

early purple orchid

lesser celandi

grey squirrel

larch

Scots pine

silver birch

Norway spruce

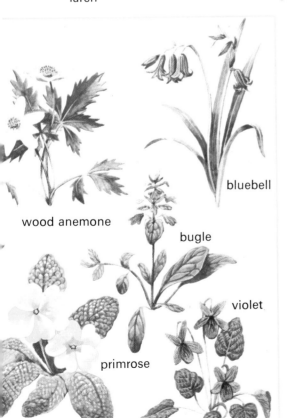
wood anemone

bluebell

bugle

violet

primrose

conifers' leaves quickly start to make food. Sticky resin in the leaves and trunk seals off any damage. The resin also stops dead leaves from decaying quickly. So the floor of a coniferous forest is thick with leaves that have not rotted into food for other plants.

Grass and shrubby plants like heathers grow beneath conifers. But far more plants grow under deciduous trees.

Deciduous trees include ash, beech, elm and oak. These trees prefer slightly warmer climates than coniferous forests. Also they have broad leaves and shed them in the autumn. This protects the trees from winter damage caused by frost, cold winds and snow. In most deciduous woods, the fallen leaves rot readily into foods for many kinds of shrub and soft-stemmed flowering plants.

An oak wood, for example, has several plant layers. Beneath the oaks is a shrub layer of hazel, hawthorn and holly. The field or herb layer spreads below the shrubs. It contains ferns and soft-stemmed flowering plants like the bluebell, primrose, wood anemone and violet. Under the field layer is the lowest layer of all. This is made of mosses and lichens that hug the ground. In late summer and autumn, these are joined by many kinds of fungi.

39

Doing Without Water

All plants need water, but some need less than others. Certain flowering plants, like cactuses, can live in very dry regions.

Some deserts are so salty that no plants can grow. Some deserts have shifting sands where no plants can put down roots. Many deserts are simply too dry for most plants. Yet deserts are often rich in the minerals plants need. In these areas, ordinary plants are plentiful where there is water near the surface.

Desert plants are mainly plants with special ways of gathering and storing water when it rains. Because rain does not sink far down in desert soil, long-lived desert plants grow some roots that spread far out just below the surface. Many plants have thick leaves with a waxy coat that prevents water loss. Others have hairs that protect the leaves from drying winds. Cactuses store water in their fleshy stems. The giant saguaro cactus of America can hold more than a barrel of water in its tall, thick stem. Cactus leaves are just spines. They lose little water and guard the stems.

Some long-lived desert plants lack waterproof stems or leaves. They survive droughts by storing water in underground roots, bulbs and tubers. Desert shrubs and trees shed leaves in drought to prevent water loss. Most desert trees also draw up water from great depths. The mesquite's roots reach down 30 metres (100 ft).

When rain falls heavily, the seeds of annual plants burst out in a mass of flowers. For a few days the desert sands are carpeted with daisies, dandelions and other soft-stemmed plants. These all wither when the drought sets in again. But the seeds they produce live on, ready to spring up and flower another year.

Right: A giant saguaro cactus from the deserts of Arizona in North America. Saguaros can live for hundreds of years.

Below: Desert plants have different ways of storing water. From left to right: *Mesembryanthemum* stores water in its thick, juicy leaves. The yucca's stiff, slender leaves reduce water loss. Small leaves also save water in the creosote bush. Spines guard the prickly pear's large swollen stems. The living stone and *Stapelia* both store water in their thick stems and leaves. The barrel cactus (shown cut open) stores water in its fat stem. All these desert plants come from America or Africa.

41

The Seed Story

Nearly all plants spend their lives producing seeds. Seeds are very important: they are the beginnings of new plants.

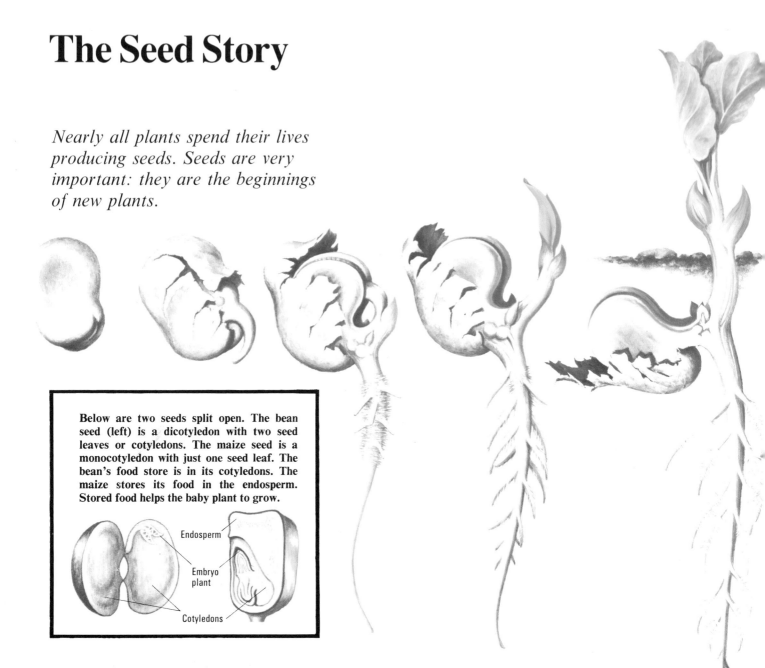

Below are two seeds split open. The bean seed (left) is a dicotyledon with two seed leaves or cotyledons. The maize seed is a monocotyledon with just one seed leaf. The bean's food store is in its cotyledons. The maize stores its food in the endosperm. Stored food helps the baby plant to grow.

Endosperm

Embryo plant

Cotyledons

If ripe seeds drop down to the ground beside their parent plant, there is not enough room for them to grow. To give them a real chance of developing into strong new plants, seeds must be spread out or scattered. Seeds and spores are scattered in four main ways. They are shot out by the parent plant, or carried away by wind, water or animals.

Many plants are specially built to help them spread their seeds. Some, for instance, are designed to use the wind. The seed capsule of a poppy has holes like a pepper pot. As the wind blows, the poppy stalk waves to and fro and the seeds are thrown out through the holes. The seeds of a willow herb have tiny parachutes that float away with the wind. Many pine seeds have wings to help them spin through the air far from the parent tree.

Other plants rely on water to take their seeds away. Coconuts and mangroves drop their seeds into the sea. The seeds sometimes drift hundreds of kilometres before they are washed ashore and start to grow. In the same way, currents carry water-lily seeds along rivers.

Certain plants have springs and catapults to help them

Above: Five steps in the germination of a broad bean seed. First a root appears. Then the shoot bursts out between the seed leaves. Shoot and root grow longer. Seed leaves stay underground and soon wither.

Below: Dandelion flowers produce seeds with little parachutes. The wind blows the seeds away and so helps dandelion plants to spread.

fire seeds or spores like bullets from a gun. Bean and pea pods twist as they ripen. The strain bursts the pods open and the seeds shoot out. The touch-me-not and other balsams have capsules with five valves. When these dry they spring open and the seeds inside fly out. Even simple plants have spore capsules that burst suddenly and catapult the spores into the air. The squirting cucumber uses water pressure to force out its seeds.

Many plants grow seeds inside a fruit. The fruit protects the seeds and helps to scatter them. Some dry fruits have hooks and prickles that cling to the fur of passing animals. Many fleshy fruits, like cherries or blackberries, have bright-coloured skins, or even black or white ones. These show up well against green leaves. Birds and other creatures are attracted by the colours and eat the fruits. But the actual seeds are not digested and they pass through the animal unharmed.

Seeds do not sprout as soon as they fall on the ground. Most of them need a rest period before growing. While they are resting, or dormant, seeds sometimes dry out completely and yet they stay alive. Many seeds lie dormant for some months before they start to germinate, or grow.

No seed germinates unless the place where it has fallen is suitable for growth. Seeds must have moisture, warmth and oxygen.

A germinating seed begins by taking in water. The seed swells and its seed coat softens and splits. Then the baby plant inside the seed bursts out. First a root grows down into the soil. Next, a tiny shoot starts pushing upward. In grasses, a sheath protects the shoot tip and it grows straight up. In most other plants there is no sheath. Instead, the shoot starts life bent over like a hook.

The young plant cannot make its own food until the shoot has grown above ground and produced leaves. Before that, the baby plant is fed by the seed. Seeds store food either in the seed leaves, known as cotyledons, or in special tissue called endosperm. The food store contains starch, sugar and fat. When the leaves grow, the cotyledons are not needed and shrivel up.

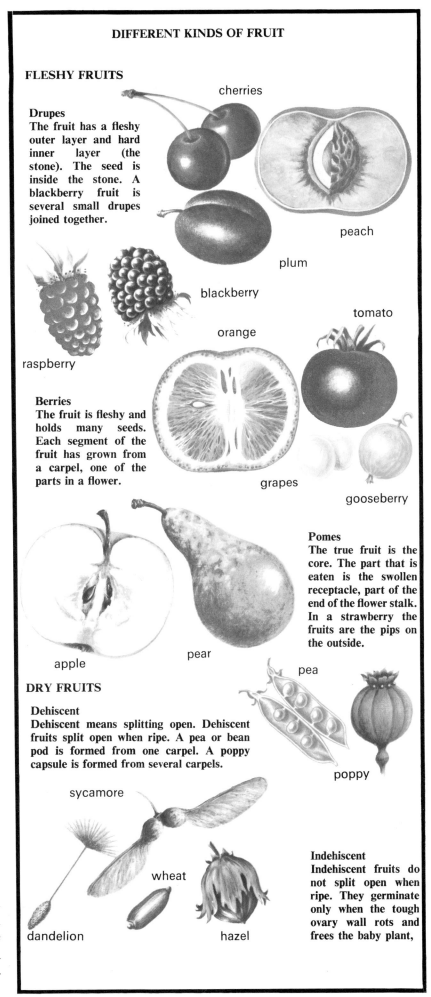

DIFFERENT KINDS OF FRUIT

FLESHY FRUITS

Drupes
The fruit has a fleshy outer layer and hard inner layer (the stone). The seed is inside the stone. A blackberry fruit is several small drupes joined together.

cherries

peach

plum

blackberry

tomato

orange

raspberry

Berries
The fruit is fleshy and holds many seeds. Each segment of the fruit has grown from a carpel, one of the parts in a flower.

grapes

gooseberry

Pomes
The true fruit is the core. The part that is eaten is the swollen receptacle, part of the end of the flower stalk. In a strawberry the fruits are the pips on the outside.

apple

pear

pea

DRY FRUITS

Dehiscent
Dehiscent means splitting open. Dehiscent fruits split open when ripe. A pea or bean pod is formed from one carpel. A poppy capsule is formed from several carpels.

poppy

sycamore

wheat

Indehiscent
Indehiscent fruits do not split open when ripe. They germinate only when the tough ovary wall rots and frees the baby plant,

dandelion

hazel

A Six-Legged World

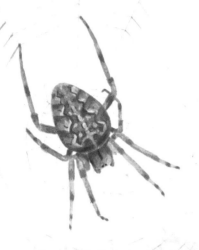

More than nine out of ten kinds of animals alive today are insects. Ants, bees, beetles, butterflies, grasshoppers and flies all belong to this great group of tiny creatures.

All insects are built on the same plan. Each one has six legs, a pair of feelers, called antennae, and a body made up of three parts: the head, thorax and abdomen. Most insects also have wings, but silverfish and bristletails have none. Unlike fishes, birds and mammals, insects have no skeleton inside their bodies. Instead, they have a hard covering outside the body. This outer skeleton provides the insect with a firm base for fixing its muscles. It also stops the creature's body from

a hoverfly trapped in
the web of a garden spider

bumblebee

damselfly

grasshopper

two ants 'talking'
to each other

plant-eating bug

drying. But the outer covering does not grow with the insect. So, from time to time, a developing insect splits its skin and steps out with a new, larger one. Some young insects develop into adults without changing shape. They are called nymphs. But many go through two big shape changes known as metamorphosis. Young insects that change shape as they grow are called larvae. A caterpillar, for example, is the larva of a moth or butterfly. After a time, the growing caterpillar forms a hard case round its body and stays still. It is then known as a pupa. One day, the case splits open and out crawls an adult moth or butterfly complete with wings.

Some insects eat plants. Others eat animals. In turn, many animals feed on insects. Among their greatest enemies are spiders. Like insects, spiders are arthropods, animals with joined legs. But unlike insects, spiders have eight legs instead of six, and bodies with two main parts instead of three.

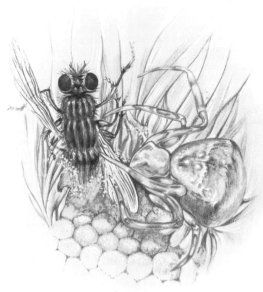

This fly has been caught by a spider. Spiders attack insects using poisoned fangs that crush or paralyze. They also trap insects in their sticky webs. Spiders are the worst enemies of insects.

monarch butterfly

ladybirds

male stag beetles use their antlers to fight about females

caterpillar

Beetles

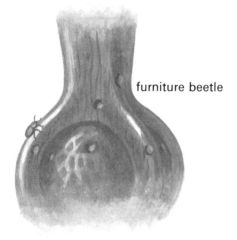
furniture beetle

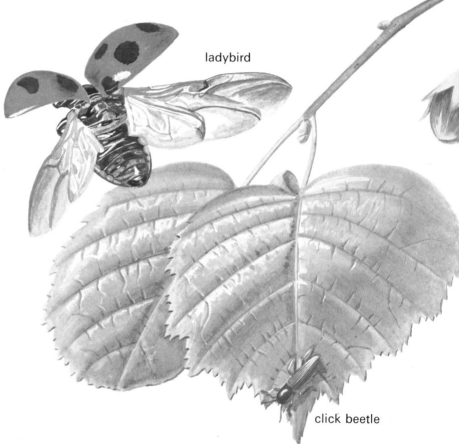
ladybird

click beetle

There are over 270,000 kinds of beetle, more than any other kind of insect. Some beetles are several inches long; others are tiny.

dor beetle

Furniture beetles damage wood furniture. The female beetle lays her eggs on wood. When they hatch, the larvae eat tunnels into the wood. The larvae are called woodworms.

Most beetles cannot bend their bodies. If they fall over they cannot easily get up. But a click beetle can. By arching its back it releases a peg under the body. The peg acts as a spring. The beetle is flicked up into the air with a clicking sound and lands the right way up.

peg

The scarab beetle rolls dung into balls and lays its eggs in them. In Ancient Egypt, scarabs were sacred; they symbolised life after death.

Dor beetles dig holes under cow dung and bury it as food. But they often bury more than they need. In this way, dor beetles help to fertilize the soil.

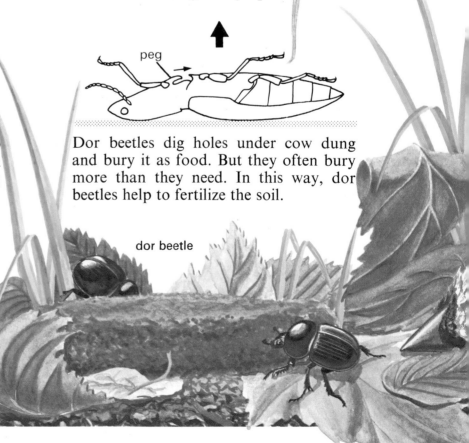
dor beetle

Weevils use their long snouts to bore into plants. The nut weevil makes holes in nuts and lays an egg in each one. The larvae feed on the kernel of the nut.

t weevil

There are about 5000 kinds of ladybird. Most of these small beetles are red or yellow with black spots. Different kinds of ladybird have different numbers of spots. Many ladybirds eat aphids, whiteflies, mealy bugs and other tiny insects. But some eat pollen, leaves or dung. Ladybirds hibernate in winter. Sometimes hundreds hibernate together under bark or in some other safe, dark hiding place.

Deathwatch beetles live in the wooden beams and boards of old buildings. They make an odd ticking sound by knocking their head against the wood to call their mate. The females lay eggs in cracks in the wood. When the larvae hatch, they burrow into the wood and stay there until they are adult. This tunnelling destroys the wood and so does great damage.

deathwatch beetle

ladybirds

The burying beetle digs holes and buries dead animals. As these may be larger than the beetle, it cannot pull them into its hole. Instead, the beetle digs away the soil from underneath so that the body sinks down. The female lays her eggs near the corpse. When the eggs hatch, the larvae feed on it.

burying beetles

Devil's coach horse

The devil's coach horse is a fierce-looking beetle. It frightens off its enemies by raising its tail and producing an unpleasant smell.

47

Insect Athletes

Grasshoppers and crickets are jumping insects. They live in fields and meadows and feed on green plants. Some kinds eat crops and can do great damage. Crickets have long feelers. Grasshoppers have short ones.

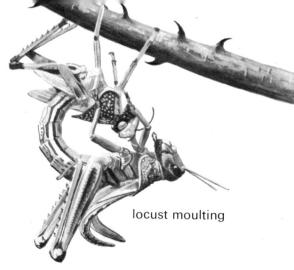

locust moulting

Locusts are large kinds of grasshopper. A young locust grows inside its skin. When the skin gets tight it splits and the nymph steps out. This happens several times before the locust is fully grown.

grasshopper

A male grasshopper chirps or sings to attract a mate. He makes the sound by rubbing the insides of his back legs against hard ridges on his wings. Most grasshoppers have large wings and can fly.

bush cricket nymph

In autumn, a female bush cricket makes a hole in the ground. She digs with the strong, sharp tube at the end of her body. She then lays eggs from the tip of the tube. The female does not survive winter. But deep down in the soil her eggs escape the cold. In the spring they hatch into grubs. Each grub quickly sheds its skin and turns into a nymph. By the middle of summer, the nymph has become an adult cricket.

great green bush cricket

Grasshoppers and crickets have long back legs that act as springs. When a grasshopper suddenly straightens its back legs it can leap twenty times the length of its own body. If the grasshopper at the bottom of the facing page made a leap it would land at the top of this page.

True crickets are related to bush crickets. Like bush crickets and grasshoppers, young crickets resemble their parents. There are many kinds of true cricket. House crickets like living in warm, indoor places. They hide by day and feed and chirp by night. House crickets eat almost any kind of plant and animal material. Male house crickets sometimes butt, bite and kick each other. But they seldom get badly hurt; the loser simply runs away.

Field crickets have no back wings and cannot fly. They live alone in burrows underground. Field crickets like hot sun and sing through the summer, day and night. When a field cricket hears footsteps coming, it crawls backwards down its hole.

The mole cricket is another burrower. This big cricket likes to dig holes in the moist banks of streams and ponds.

field cricket

house cricket nymph

house cricket

The mole cricket burrows with big, strong forelegs like a mole's. It feeds on roots but also eats worms and larvae.
mole cricket

49

Ponds and Streams

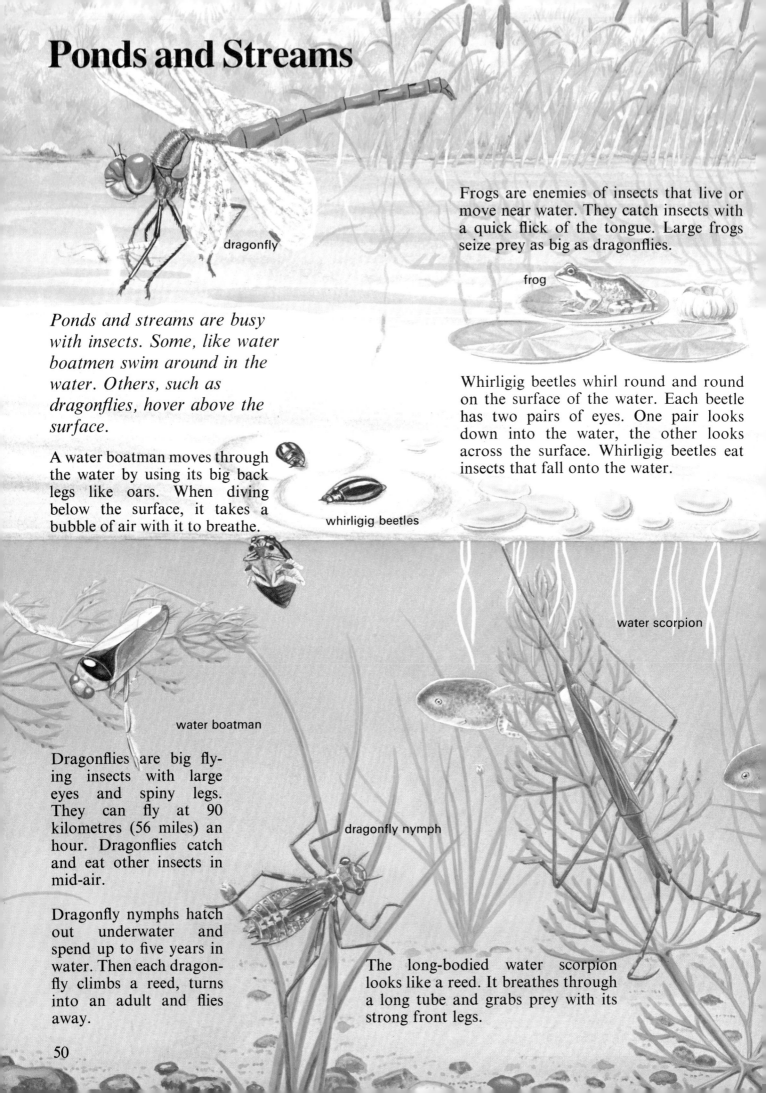

dragonfly

Frogs are enemies of insects that live or move near water. They catch insects with a quick flick of the tongue. Large frogs seize prey as big as dragonflies.

frog

Ponds and streams are busy with insects. Some, like water boatmen swim around in the water. Others, such as dragonflies, hover above the surface.

A water boatman moves through the water by using its big back legs like oars. When diving below the surface, it takes a bubble of air with it to breathe.

whirligig beetles

Whirligig beetles whirl round and round on the surface of the water. Each beetle has two pairs of eyes. One pair looks down into the water, the other looks across the surface. Whirligig beetles eat insects that fall onto the water.

water scorpion

water boatman

Dragonflies are big flying insects with large eyes and spiny legs. They can fly at 90 kilometres (56 miles) an hour. Dragonflies catch and eat other insects in mid-air.

dragonfly nymph

Dragonfly nymphs hatch out underwater and spend up to five years in water. Then each dragonfly climbs a reed, turns into an adult and flies away.

The long-bodied water scorpion looks like a reed. It breathes through a long tube and grabs prey with its strong front legs.

50

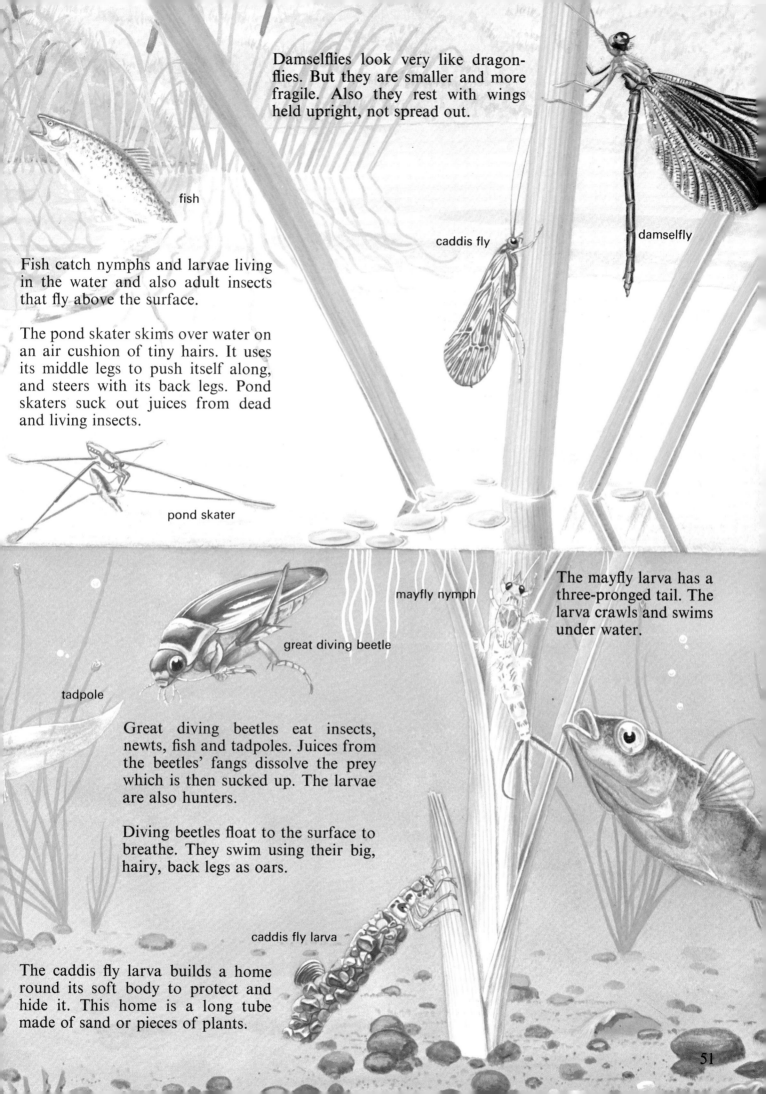

Damselflies look very like dragon-flies. But they are smaller and more fragile. Also they rest with wings held upright, not spread out.

fish

caddis fly

damselfly

Fish catch nymphs and larvae living in the water and also adult insects that fly above the surface.

The pond skater skims over water on an air cushion of tiny hairs. It uses its middle legs to push itself along, and steers with its back legs. Pond skaters suck out juices from dead and living insects.

pond skater

mayfly nymph

The mayfly larva has a three-pronged tail. The larva crawls and swims under water.

great diving beetle

tadpole

Great diving beetles eat insects, newts, fish and tadpoles. Juices from the beetles' fangs dissolve the prey which is then sucked up. The larvae are also hunters.

Diving beetles float to the surface to breathe. They swim using their big, hairy, back legs as oars.

caddis fly larva

The caddis fly larva builds a home round its soft body to protect and hide it. This home is a long tube made of sand or pieces of plants.

Spiders and Ants at Home

Creatures of all kinds make fascinating homes. Two expert builders are spiders and ants. Spiders weave delicate webs out of silk. Many species of ant, working together in a group, construct remarkable mansions under the ground.

garden sp

Above: The spider clings to its web with special hooks on its feet. Right: The tip of a spider's foot, showing the hook. This picture is greatly enlarged.

hook

SPIDER'S FOOT

Below: A trapdoor spider springs from the door of its underground home to capture and kill a cricket.

Below: Another trapdoor spider eats a millipede it has pulled into its tube.

The garden spider makes a home of silk threads. These threads come from special parts of its body called spinnerets. The spider, using its legs, spins the threads into a sticky web. Then it hangs near the middle of the web and waits for its victims. As insects fly into the web they become trapped. The spider then grabs and eats them.

The trapdoor spider has a tube-like nest under ground. To build its home, the spider digs a tunnel then lines it with silk threads. It covers the nest's entrance with a strong lid made of silk and soil. The lid is fastened to the ground by a silk hinge. The spider waits for prey just inside the half-open trapdoor. As soon as an insect or some other small creature comes close, the spider pounces. It bites its victim to paralyze it then pulls it down into the nest. The trapdoor slams shut behind them.

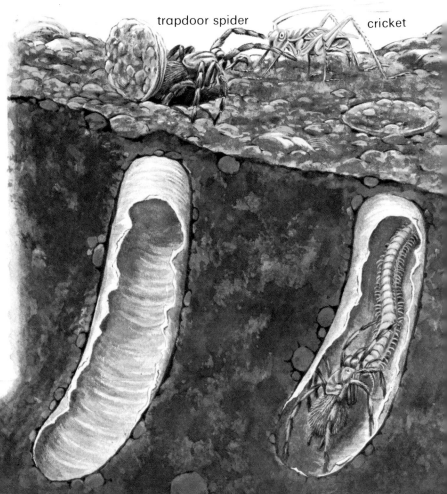

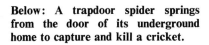

trapdoor spider   cricket

Weaver ants live in trees and make nests from leaves. Several worker ants pull the edges of the leaves together. Other workers bring ant larvae and press the mouths of the larvae against the leaf edges. The larvae produce silk threads to join the leaves together. Weaver ants belong to the tropical forests of Asia.

larva

weaver ant nest

silk threads

anthill

larvae

eggs

Most ants make their homes in the ground. In some cases, a group or colony of ants piles up soil into an anthill. Inside, the ants dig out little rooms or chambers. There are many different chambers: nursery chambers for the eggs, others for larvae and pupae, storerooms for keeping food and special rooms for rubbish. Each ant has a certain job such as builder, nurse or soldier. An extra large ant, called the queen, lays all the eggs.

Below: Three kinds of termite nest. One nest is cut open to show the inside. The anteater raiding a tree nest catches termites on its sticky tongue.

anteater

Termites are ant-like insects that live together in colonies. Some termites build nests under the ground. Others build great mounds as high as houses and as hard as concrete. A few termites nest in trees. In each colony the large queen termite lays thousands of eggs every day. Worker and soldier termites are blind. Soldiers have large heads with huge, strong jaws. They protect the nest from enemies.

worker

soldier

king

queen's body swollen with eggs

53

Insects On Guard

Insects escape from enemies in different ways. Some hide; others look very fierce.

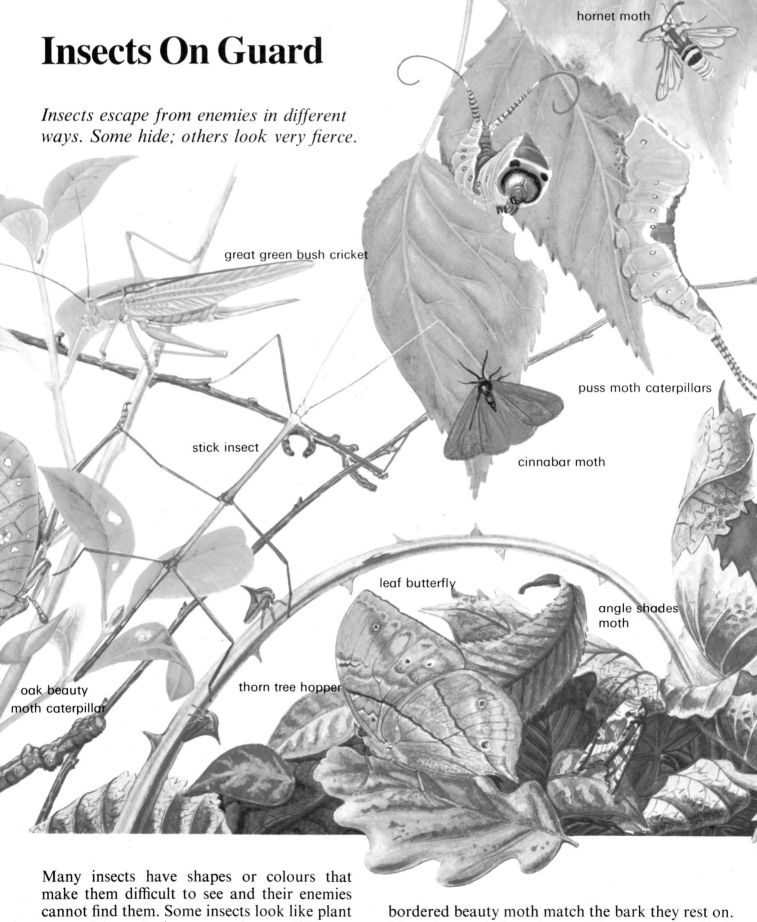

hornet moth

great green bush cricket

puss moth caterpillars

stick insect

cinnabar moth

leaf butterfly

angle shades moth

oak beauty moth caterpillar

thorn tree hopper

Many insects have shapes or colours that make them difficult to see and their enemies cannot find them. Some insects look like plant stems. The oak beauty moth caterpillar resembles a brown, knobbly twig. The stick insect also looks like a twig, but a long, thin one. Thorn tree hoppers are easily mistaken for sharp thorns on a branch. The colours of the goat moth and the pattern of the dark bordered beauty moth match the bark they rest on. Many insects imitate leaves. The great green bush cricket clinging to a branch looks like a folded leaf. The long thin body of the pine beauty moth caterpillar is striped like the long thin pine needles that it lives among. The angle shades moth and the leaf butterfly both have colours that make them

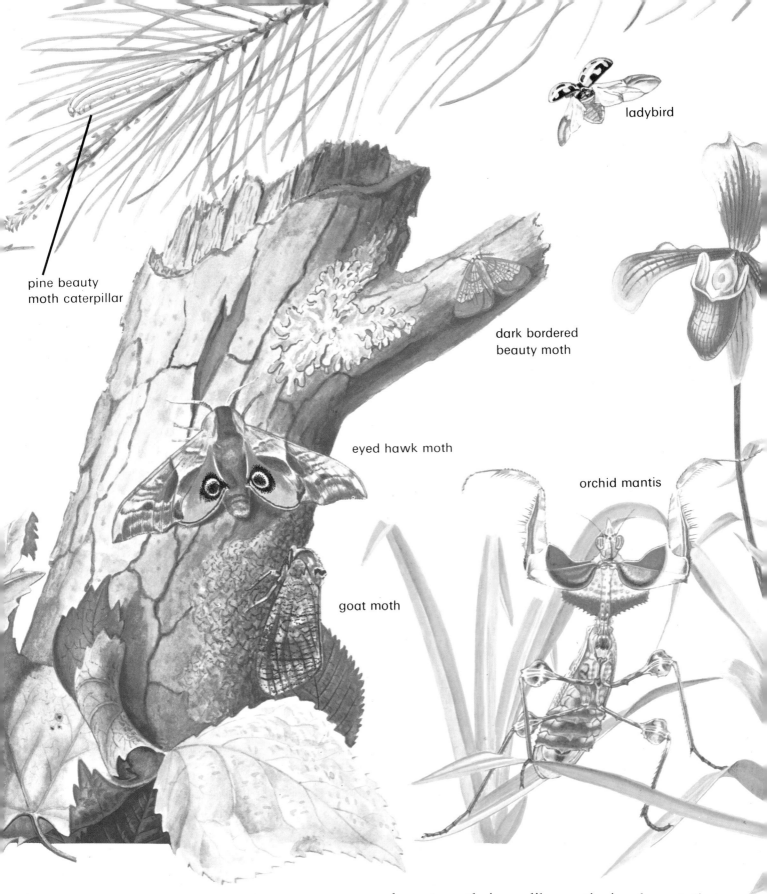

pine beauty
moth caterpillar

ladybird

dark bordered
beauty moth

eyed hawk moth

orchid mantis

goat moth

look more like dead leaves than insects.

The orchid mantis is disguised as a flower to attract its prey and fool its enemies.

Certain insects do not need to hide. The cinnabar moth and ladybird have bright colours as reminders of their bitter taste. Yet other insects rely on bluff to scare off enemies. The hornet moth is so like a stinging hornet that creatures leave it alone. The great spots on an eyed hawk moth seem to be owl's eyes to its enemies. When threatened, the puss moth caterpillar hunches its shoulders so that the front of its body becomes a frightening face. This caterpillar also spins red threads from its tail and squirts out a smelly liquid.

55

Friends and Enemies

Man has both friends, and enemies, in the insect world.

Insects may be friends for several reasons. Some insects help to clear pests and weeds from farms and gardens. Once, ladybirds were taken from Australia to America to help kill off an insect that was destroying the orange trees. The ladybirds managed to save the orange crop. Dragonflies eat mosquitoes. The ichneumon fly often lays eggs on caterpillars that harm crops. When the ichneumon larvae hatch out, they eat and kill the caterpillars.

Of course, dragonflies and ladybirds choose their prey by instinct. They do not know they are killing animals that man calls pests. (In fact, they also kill creatures that man calls friends.)

Other insects, especially bees and butterflies, are friends because they help man grow food. They do this by carrying pollen from plant to plant which fertilizes flowers. From flowers come all the fruits and seeds that people eat.

Some insects destroy weeds. The cactus moth was taken to Australia so that its larvae could eat and kill the prickly pear cactus. This cactus was destroying the pasture lands.

A few insects produce useful substances. Honeybees make honey. Cochineal, a red food colouring, comes from a Mexican insect. The lac insect, of Asia, gives out a sticky liquid which is used to make wood varnish.

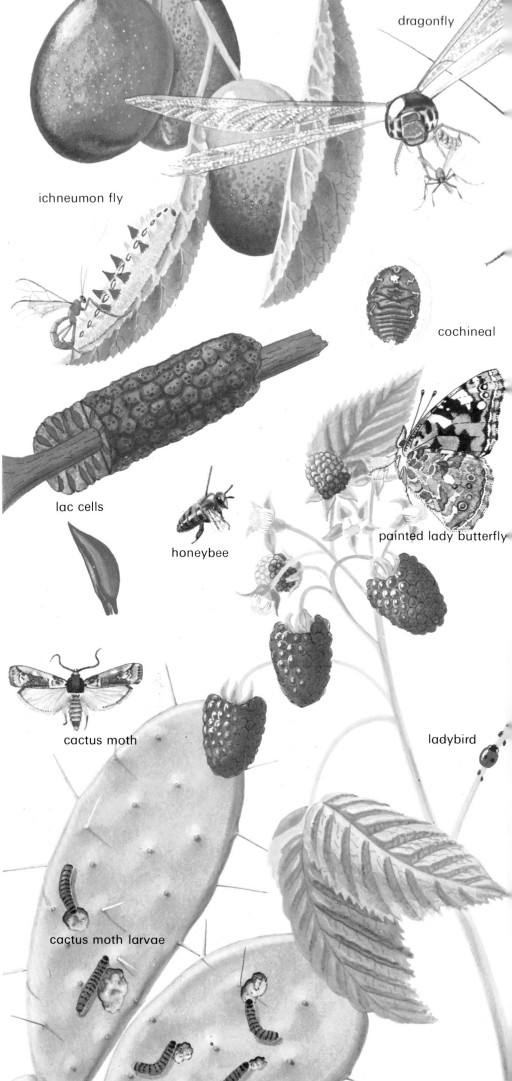

dragonfly

ichneumon fly

cochineal

lac cells

honeybee

painted lady butterfly

cactus moth

ladybird

cactus moth larvae

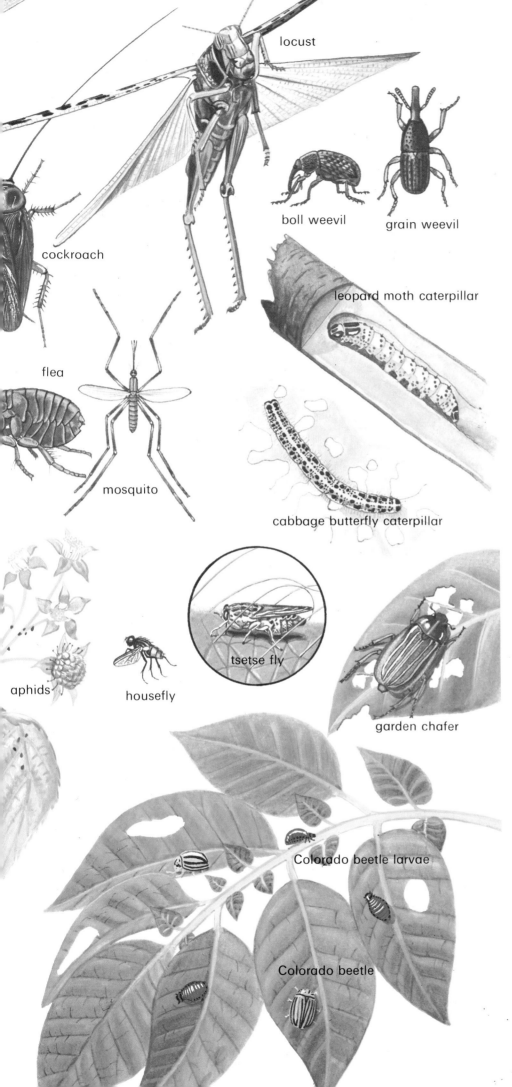

locust

boll weevil

grain weevil

cockroach

leopard moth caterpillar

flea

mosquito

cabbage butterfly caterpillar

aphids

housefly

tsetse fly

garden chafer

Colorado beetle larvae

Colorado beetle

But many insects are man's enemies. Some are parasites that grow and feed on other creatures and so spread various diseases. For example, fleas live on the blood of people and animals. A flea's bite is very irritating. Fleas can carry diseases from animals to man.

Some mosquitoes carry the germs that cause malaria. This fever kills many people in Asia and Africa every year.

The bite of the tsetse fly gives sleeping sickness to its victims. In parts of Africa this disease has killed thousands of people and has made it impossible to raise cattle and horses on farms.

Even the common housefly spreads disease. This insect moves from filth to food with germs on its mouth and feet. The common housefly can carry up to thirty different diseases. The cockroach is another of man's insect enemies. It spreads dirt and disease throughout the house.

Some kinds of beetle and moth are pests because they eat and spoil stored food. Grain weevils, for example, attack grain that has been harvested and placed in store.

Other insects are enemies because they feed off living animals and plants. Greenflies and other aphids suck plant juices. Colorado beetles eat and destroy potato plants. Boll weevils attack the fluffy seed-pods of the cotton plant. Garden chafers damage the leaves and flowers of many crops and trees. The leopard moth caterpillar burrows into tree trunks. The cabbage butterfly caterpillar eats cabbage leaves.

But few insects are as terrible as locusts. When a locust swarm settles upon a field, the insects eat every green thing they can see.

57

How Amazing!

The insect world has many strange members. Some insects just look weird. Others have very peculiar habits.

The longhorn beetle is one of the largest beetles alive. Some have feelers four times as long as their body. Some have bodies as big as a man's hand.

The thorn moth caterpillar has no legs under the middle of its body. It walks by moving first its front legs, then its back legs. As it brings its back legs forward, its body bends and forms a loop.

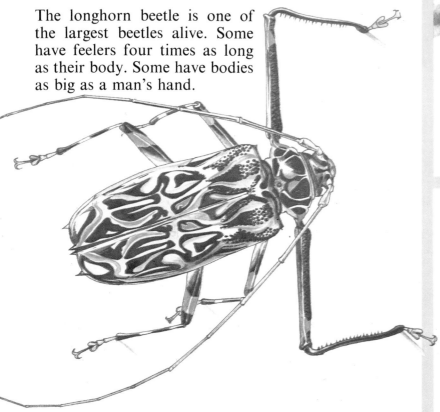

The female earwig is a surprisingly good mother. Most insects lay eggs and leave them. The earwig looks after hers. She licks them and moves them about to stop them going moldy. When the young hatch, she brings them food. Later, outside the nest, she shows them where to find food. The mother earwig and her brood stay together until the young are fully grown. The common earwig lives in many parts of the world.

In summer, there are often frothy blobs of cuckoo spit on plants. Each blob is made by the nymph of the froghopper. This young plant-sucking insect surrounds itself with froth to prevent its body drying up and to protect it from enemies. Adult froghoppers escape by jumping.

The tiny flea (shown greatly magnified) can jump 110 times its own body length, and 130 times its height.

Millions of years ago this insect was trapped in the sticky resin of a pine tree. In time the resin turned into hard, yellowish amber. The dead insect's body still lies inside it. Amber preserves many insects that died out long ago.

Hornets and wasps build paper nests in hollow trees. They make the paper by chewing wood into a paste. Sometimes they mix paper with soil. This tree is cut open to show the nest cells inside.

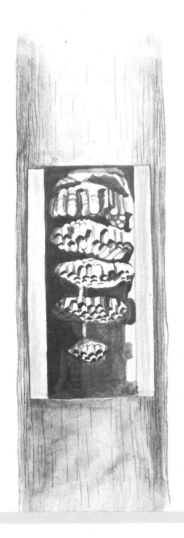

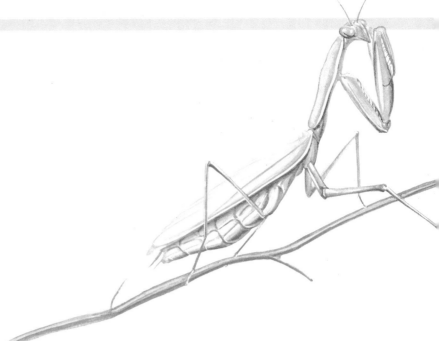

The praying mantis holds up its front legs like someone praying. But these spiny legs are weapons. The mantis uses them to grab insects and other prey.

Honeypot ants live in the desert where food is often scarce. When flowers bloom, worker ants collect the nectar. They feed this to other ants until their bodies bulge with food. These ants grow too fat to move and become living larders for the ant colony.

Some termite mounds are three times taller than a man and contain millions of termites.

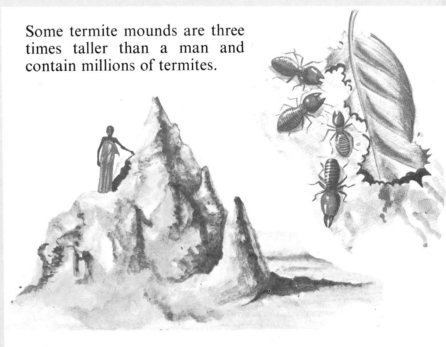

Nests, Eggs and Chicks

Birds usually give special care to their eggs and young chicks.

In the breeding season birds need somewhere safe to lay and hatch their eggs and raise their chicks. Many birds build nests high up in trees. Others build on the ground or even in burrows.

Most birds build nests of mud or of twigs, grass, moss or feathers. There are birds that lay eggs on bare rock or stones and birds that bury their eggs in warm heaps of dead and rotting plants.

Eggs hatch only if they are kept warm. Most birds keep their eggs warm by sitting on them, generally for several weeks. In some cases, just one parent sits. In others, the two parents take turns.

When the chicks hatch, many of them are quite helpless. They often have no feathers and their eyes are shut. Both parents usually feed and protect them. Most young birds develop quickly and learn to fly and feed themselves.

A herring gull chick pecks, by instinct, at the red spot on its parent's beak. This makes the parent bring up swallowed food and gives the chick a meal.

The tailor bird (1) sews leaves together in a cup, then builds his nest inside. Weaver birds (2) make nests like hanging baskets. The oven bird (3) shapes a mud nest that looks like an old-fashioned oven. The fairy tern (4) just lays her eggs on a branch.

young herring gulls

reed warbler

young cuckoo

female mallard duck

ducklings

Above: European cuckoos take no interest at all in their young. They lay their eggs in the nests of other birds. The chicks are then hatched and raised by the foster parents. Each female cuckoo lays about 12 eggs. She lays them, one by one, in different nests. When she puts her own egg in a nest, she takes out an egg belonging to the host bird. The young cuckoo hatches quickly. It is a large creature and soon throws out its foster parents' eggs or chicks. With no young of their own, the foster parents spend all their time feeding and looking after the baby cuckoo.

Raising chicks often means hard work for the parent birds. From dawn to dusk they make non-stop trips in search of insects, worms, seeds and other food. Chicks eat half their body weight in food every day. A pair of blue tits, for instance, bring about ten thousand caterpillars to a brood of ten young tits in only eighteen days. After that time most young birds can fly, but they still cannot feed themselves. Yet, in spite of all this hard work, some kinds of birds raise three broods in one year.

Not all parent birds have to spend so much time finding food for their young. Some chicks can look after themselves almost at once. Young pheasants and partridges are soon able to run about. Young ducklings can swim. Such birds find food for themselves. But at first they follow their mother for protection.

Often, chicks are covered in fluffy grey-brown feathers. These colours make the young bird match its nest and it is then harder for enemies to spot it. The parents also use their own brighter colours to distract an enemy and draw it away from the nest.

Although parent birds put so much work and effort into looking after their chicks, many young birds die of starvation or are killed by enemies. But enough survive to keep the bird population steady.

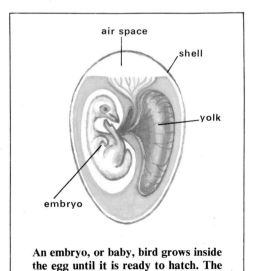

air space

shell

yolk

embryo

An embryo, or baby, bird grows inside the egg until it is ready to hatch. The tiny creature breathes air from a space at the end of the egg and gets its food from the yellow yolk.

swallow's nest and young

61

Following the Sun

Some birds always leave winter behind. Every year, when the days start growing shorter and colder, they set out on a great journey. They follow the sun in search of warmth and food.

Many birds have two homes. They lay eggs and raise their young in northern lands during the summer. These birds find plenty of food in the long hours of daylight. But, as summer ends, the days grow shorter and colder and food becomes scarce. Then, the birds fly to warmer parts of the world for winter. In this way, they are sure of finding food all the year round.

Birds that travel between a winter and a summer home are called migrants. Many kinds of birds migrate. Swans, storks and eagles are the largest migrants. Hummingbirds, wrens and warblers are the smallest. Most migrants are birds from cool lands. In hot lands there is not much difference between summer and winter, so tropical birds do not need to change homes. Even in cold parts of the world there are a few birds that stay through the winter. Some of them, like tits, do this by changing their diet. In summer, they live on insects but eat seeds and berries in winter.

Many migrants fly thousands of kilometres each year. Perhaps the greatest traveller is the Arctic tern. This seabird spends the northern summer in the Arctic. But before the polar winter sets in, the tern flies half way round the world to spend the southern summer in the Antarctic.

Migrating birds fly in flocks. Even birds that usually live alone, form flocks to migrate. By flying together they stand a better chance of surviving the long journey. Even so, many die in storms or drop from exhaustion.

swallow

EUROPE

a bullfinch watches
the swallows depart

Swallows are sun-loving birds. They arrive in northern parts of the world in late spring. There, they build nests, lay eggs and raise their young during the summer. In these warm months, there are plenty of flying insects for them to eat as they swoop and glide through the air. But when winter comes, most flying insects die and there is no food left for the swallows.

62

Swallow

Arctic Tern

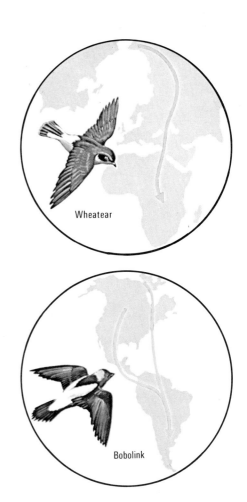

Wheatear

Bobolink

The migration routes of four long-distance travellers: the swallow, Arctic tern, wheatear and bobolink. The Arctic tern covers the greatest distance. At the end of the northern summer it sets off on its 160,000 km (10,000 mile) flight from the Arctic to the Antarctic. Six months later, the tern flies back again. Wheatears are almost as remarkable. They nest in Europe, Russia and Alaska. Those from Alaska have an incredible journey. They fly over the Bering Straits and then across Asia to Africa. In the Americas, the bobolink travels from its breeding grounds in the north to the grasslands of the south.

These migration routes are worked out by ornithologists (scientists who study birds). They capture birds and put small rings on their legs. Later, if the birds are re-captured, the scientists can record their travels. Radar is also used to trace the movement of flocks.

One of the greatest mysteries of the animal world is how migrating birds find their way across oceans and continents. Some of them do so with great precision as every year they go to exactly the same spot.

Birds migrate by instinct. Young birds, migrating for the first time, can find their way thousands of kilometres over land and sea without any help from their parents. But scientists have discovered that migrants also make use of signposts along the way. Somehow, they use the sun and stars, wind currents and landmarks such as rivers to guide them.

Not all migrants travel across entire continents. Some Arctic birds leave the icy north in winter but only go as far as Britain or France. In mountain lands, migrations can be far shorter. Birds that spend the summer on the high slopes often shelter in valleys in winter.

Another migration mystery is timing. Some birds clearly act by the weather. When the temperature drops they leave. Others wait until the wind is right to speed them on their way. But there are some, like swifts, that migrate on exactly the same date every year, as if they had been programmed by a computer.

So, as the days become shorter and cooler, the swallows gather in flocks and set off on the long flight south. Their journey takes them across the Mediterranean Sea and the Sahara Desert to the southern part of Africa. The swallows live for a few months on the warm plains and then begin the long trip back to the north to breed.

AFRICA

a carmine bee-eater watches the swallows arrive

Finding Food

Some birds eat seeds. Some eat insects. Or fish. Or worms.

Many birds have special ways of eating their kind of food.
Birds often have a beak specially shaped for certain foods. The hawfinch's thick, strong bill cracks open hard shells. The hummingbird sucks nectar through its tube beak. The hedge sparrow's sharp beak seizes insects. The darter uses its long pointed bill for spearing fish. The pelican's pouched beak acts as a fishing net. Eagles have strong, hooked beaks for tearing fish. The zig-zag edges of a merganser's beak help it grip fish. The flamingo's scoop beak strains food from water.

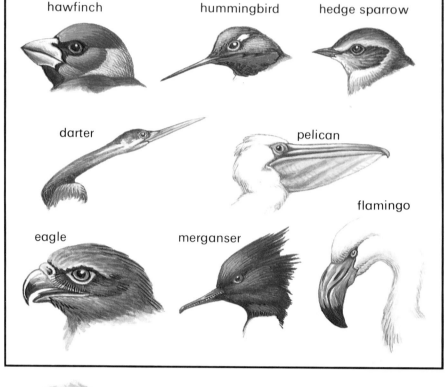

A woodpecker finch digs insects out of bark using a spine taken from a cactus.

The Egyptian vulture drops a stone on an ostrich egg to break it open.

A song thrush beats a snail against a stone to smash open its shell.

A pigeon can suck up water with its head bent down. But a sparrow takes a beakful and then throws back its head to swallow. Most birds have to drink in this awkward way.

Some birds have unusual ways of obtaining food. The pictures on the left show three birds using different tools to help them get at their food. There are also birds that make use of other animals to help them find something to eat. Skuas, for instance, steal fish caught by other seabirds. The gulls below are seizing worms turned up by a plow. Sparrows steal, beg and scavenge food produced by man.

Keeping Clean

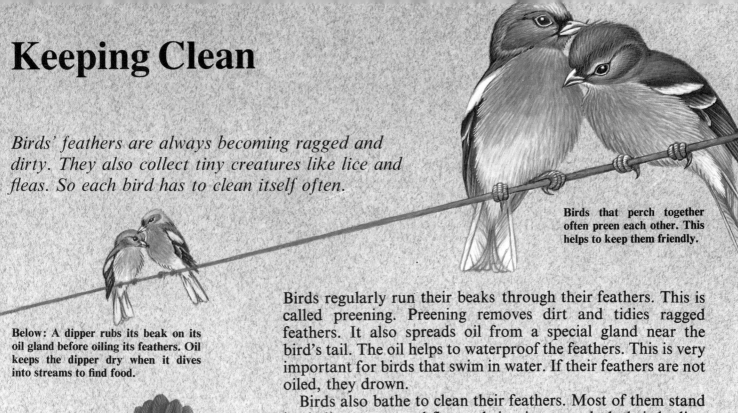

Birds' feathers are always becoming ragged and dirty. They also collect tiny creatures like lice and fleas. So each bird has to clean itself often.

Birds that perch together often preen each other. This helps to keep them friendly.

Below: A dipper rubs its beak on its oil gland before oiling its feathers. Oil keeps the dipper dry when it dives into streams to find food.

dipper

Birds regularly run their beaks through their feathers. This is called preening. Preening removes dirt and tidies ragged feathers. It also spreads oil from a special gland near the bird's tail. The oil helps to waterproof the feathers. This is very important for birds that swim in water. If their feathers are not oiled, they drown.

Birds also bathe to clean their feathers. Most of them stand in shallow water and flutter their wings to splash their bodies. Chickens, pheasants and some others prefer a dust bath. They flap their wings to cover themselves with dust. Then they begin to preen. No bird can use its beak to preen its head. Instead, a bird scratches its head with a claw or rubs it on the oiled feathers of its body.

Even well-preened feathers do not last for ever. Each year, birds moult. They shed old feathers and grow new ones.

Below: Sparrows taking a dust bath. Sparrows also bathe in ponds and puddles.

sparrows

Below: A starling places ants among its feathers. The ants produce an acid that kills tiny pests living on the bird.

heron

greenfinches

arling

goldfinch

Above: A heron scratches its plumage with a claw to clean off muddy slime. Some of its feathers produce a powder that helps to dry the slime before the heron cleans it off. The feathers are called powder down.

Up, Up and Away

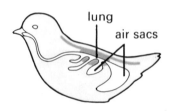

Most birds can fly. They fly by beating their wings. Many of them also glide with the wind and air currents.

Man has always admired the way that birds fly. Hundreds of years ago people tried to imitate birds. They strapped artificial wings to their arms and tried to flap like birds do. But no birdman ever flew. Man is too big and too heavy to lift his body into the air by flapping artificial wings. The largest birds are also too heavy to fly.

Birds that can fly are built in a very special way. First, they have thin, hollow bones. These help to keep their bodies light. Next, they have very powerful muscles. More than half a bird's weight is in its muscles. The largest muscles are the ones that work the wings. Lastly, bird's bodies are covered with feathers that grow from sockets in the skin. No bird could fly without feathers on its wings.

A bird's wing is curved on top like an aircraft wing. As the bird moves forward, air rushing over the wing is forced to travel farther than air rushing under the wing. So the air above is spread more thinly than the air below. The wing is pushed up by the air below into the thin air above. This gives the bird lift. Lift is the force that keeps a bird in the air.

The larger the wings or the faster the bird, the greater the lift. Fast-flying birds need smaller wings than slow flyers of the same size. Gliding seabirds gain height by turning into the wind so that it speeds past their wings.

Above: Inside a bird's body there are air sacs, rather like bubbles. The air sacs help to lighten the bird.

Below: A bird's hollow bone cut open to show thin struts that help to strengthen it.

Its big primary feathers drive a bird forward.

The secondary feathers help shape the wing to give lift.

Small feathers, called coverts, give the wing a smooth surface.

A bird uses its tail feathers to steer, brake and keep stable.

Flight muscle joins breast bone to wing.

FEATHERS

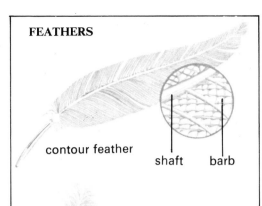

contour feather

shaft   barb

down feather

Birds have two kinds of feather. Strong contour feathers cover the outer body and wings. The circle shows part of a contour feather enlarged. Tiny kooks, or barbs, join the separate parts of the feather that sprout from the shaft. Soft down feathers lie between the skin and the contour feathers.

From left to right: A great tit in flight. As the bird takes off, it opens its wings and pulls in its legs. It then raises its wings to a full upstroke. Next, the tit pulls its wings down, pushing the air away. As the bird lifts its wings for another upstroke, the primary feathers part to pass quickly through the air. To land, the bird brakes by spreading its wings and tail.

WING SHAPES

An eagle's broad wings give it enough lift to soar high on rising air currents.

With its narrow, pointed wings, the swift flies fast but can still twist and turn.

The albatross uses its long, narrow wings to glide over the sea for days on end.

Short, broad, strong wings give a duck plenty of lift. If disturbed on land it can quickly rise into the air.

Survival Tactics

Birds have many enemies on the ground and even in the air. But they have various ways of protecting themselves.

Most birds have some defence against enemies. Many are difficult to see when they keep still because their colours match their surroundings. Brown, black and yellow feathers help to hide birds that crouch among dead leaves. White feathers camouflage birds that live in snowy lands.

nightjar

woodcock

The female woodcock's back and wings are chestnut, black and silver-grey. Dark bars run across her pale breast. These colours form patterns that make the bird nearly invisible among the dead leaves of a woodland floor. The woodcock makes her nest there in a shallow dip in the ground. Her eggs almost match the colours of the dead leaves lining the nest. Very few of the woodcock's enemies can ever find her nest.

The nightjar is grey with mottled brown markings. When the bird crouches on a dead branch, few passers-by notice it. The nightjar hides like this during the day. When dusk falls, it flies about searching for moths and other insects.

The snowy owl and the ptarmigan live in cold lands where snow covers the ground in winter. Their white feathers help to camouflage them against the snow. The ptarmigan uses its camouflage to hide from enemies like the snowy owl.

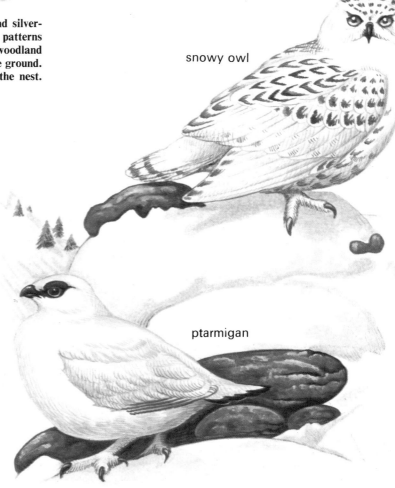
snowy owl

ptarmigan

The garden warbler has a brown back, a grey-white belly and a yellowish breast. These colours help to hide the bird when it is perching among the leafy branches of a shrub or bush even though it may be singing loudly. Other members of the warbler family are camouflaged by colours in this way.

The eggs of birds that nest on stony ground are difficult to see because their colours match the pebbles lying round them.

Many adult birds protect their eggs and young. Terns and gulls dive-bomb people walking through their colony. Tits mob a hawk if it comes near. A female killdeer pretends she has a broken wing so as to lure a prowling fox far from her nest.

Most adult birds can fly away from danger. Some fly in flocks which helps protect them from birds of prey. Rheas, ostriches, emus and cassowaries are big birds that cannot fly. But some of them can run very quickly, as quickly as a horse can gallop.

ostrich

lion

An ostrich runs away from a lion. Its long, strong legs speed the bird along at up to sixty kilometres (thirty-seven miles) an hour. An ostrich also has a very savage kick.

fox

killdeer nest

A killdeer, from North America, pretends to be hurt. She lures the fox far from her nest then flies safely away.

killdeer

69

Birds of the North

Only hardy birds survive the icy cold of winter in the northern forests. But in spring, other birds fly in from the south to find food and to nest.

The forests of the far north cover much of Canada and the northern parts of Europe and Asia. Not many birds live all year in these northern forests. Firstly, because there is not much food. The forest trees are mainly conifers. Most creatures do not like conifers. They do not like eating the resin-flavoured wood and leaves. But few other plants grow in the forest. Secondly, because winter in the far north is very long and very cold.

Yet some birds make their home in these forests. A few kinds of grouse feed on the buds and leaves of conifers. Crossbills use scissor-shaped beaks to cut through cones. A small crow called the nutcracker breaks open cones with its bill; it also eats berries, nuts and insects.

Insects are the main food of some other birds. Tits, chickadees and nuthatches somehow always find enough insects to eat, even in winter. Woodpeckers use their long, strong beaks as picks. They dig out insects that burrow in the bark of trees. Then they catch the insects on their long, barbed tongues.

There are also birds of prey. By day, hawks sweep among the trees in search of birds and mammals. At night, owls take up the hunt.

owl

woodpecker

crossbill

grouse

Southwards, the great forests of conifers give way to broadleaved woodlands. Broadleaved trees shed their leaves in winter. These woodlands are often light, leafy places with shrubs and flowers under the trees. Many birds find food and shelter here. Finches eat buds and seeds. Wrens and tits feast on tiny insects. Tree creepers hunt for insects on tree bark. In summer, insect-eating visitors join the search for food. Warblers feed among the twigs and branches. Flycatchers swoop on insects as they dart across an open glade. Swallows patrol the space above the trees.

Other birds are less particular about their diet. Thrushes feed on berries in the autumn but eat hibernating snails in winter. Jays steal eggs and young birds in spring. In winter, they largely live on acorns.

Falcons cruise among the trees. They kill small birds like finches, warblers and swallows. At night, owls hunt for mice and voles.

Left: Birds in a coniferous forest. An owl flies silently in search of prey. A woodpecker clings to a tree and hunts for insects. The crossbill opens a fir cone to eat its seeds. The grouse finds food on the forest floor. This page: Some birds of deciduous woodlands. A wren sings, perched in an oak tree. A blue jay steals an egg from a smaller bird. Tree creepers search for insects in the bark. A swallow catches insects as it flies around.

swallow

tree creepers

blue jay

wren

Birds in the Tropics

The world's tropical forests are full of birds. Some fly among the sunlit leaves of the tree tops. Others live lower down in the shade.

More kinds of birds live in tropical rain forests than in any other place. This is mainly because there are so many plants for them to eat. Hundreds of different plants thrive in the warm, damp atmosphere of a tropical forest. Huge trees with tall, bare trunks grow close together. Far above the forest floor their leaves meet to form a giant green umbrella.

Some birds find food above the tree tops. Swifts dash about in search of insects. Hawks and eagles soar above the green umbrella. From time to time these birds of prey swoop down to seize bats, birds or even monkeys from the forest's top branches.

Most birds live high up in the branches of the umbrella. There, the many flowers and fruits form food for birds like

Tropical forest birds from different countries. Toucans belong to tropical America. They use their big, but lightweight, beaks to tear off chunks of fruit. Harpy eagles also live in tropical America. They use their powerful legs to seize prey as large as monkeys. Hornbills are found in Africa and Asia. They get their name from the horny outgrowth on the top of their beak. Parakeets come mainly from Africa and Asia. This one's bright green colour helps it hide among the leaves. Turacos live in Africa. They feed on fruits and insects. Birds of paradise are chiefly found in New Guinea. Males woo females by showing off their plumage

harpy eagle

hornbills

toucan

toucans, parrots and hummingbirds. All these birds have developed special ways of managing their food. Hummingbirds hover over flowers like tiny helicopters. While they hover, they plunge their long, thin beaks into the petals to obtain a meal of nectar. Hummingbirds are the smallest birds in the world. Some are not much bigger than a bumblebee. Toucans have huge beaks shaped rather like canoes. Some have beaks almost as large as their bodies. These beaks help toucans to reach fruits that are out of reach for short-beaked birds. Hornbills find their large beaks useful, too. A parrot can crack open nuts with its powerful beak and use this as an extra claw for climbing. Parrots also have powerful grasping feet. They are the only birds that can clutch food in one foot.

In the dim light beneath the tree tops, small birds hunt insects. Babblers, bulbuls, flycatchers and tits dart among the tree trunks.

The gloomy forest floor has less food for birds than the sunlit branches high above. But jungle fowl and pheasants scratch about among dead leaves for seeds and insects. The cassowary, a large flightless bird, also lives upon the forest floor.

birds of paradise

turaco

parakeet

73

Birds of the Grasslands

Birds that eat seeds or insects find plenty of food in the tall grasses. This open country is also a rich hunting ground for birds of prey and scavengers.

In Africa, the land near the equator is covered with thick wet forest. Further to the north and south, the country is much drier and more open. It is a land of tall grasses and scattered trees. The birds living in these dry grasslands are quite different from birds found in the wet tropical forests.

Many grassland birds, like weaver birds, feed on the small dry seeds of grasses and other plants. Weaver birds are related to the house sparrow. One kind is called the quelea. This is a small, grey-brown bird with a red beak. Queleas fly over the grasslands in huge flocks. Just one flock can contain several million birds. In fact, there are probably more queleas on Earth than any other

Six different birds from Africa's grasslands. Bright-coloured bee-eaters live on insects. Each pair nests in a hole dug in a soft cliff. A stretch of cliff often contains many holes. The oxpecker in this picture is perching on the ground. But generally, oxpeckers spend their time hunting ticks on the backs of big wild mammals such as rhinoceroses. The secretary bird is about to attack a snake. This large, hawk-like bird hunts prey on the ground. Behind the secretary bird are two ostriches. The male has the smart black and white plumage. On the right, two marabou storks eat the remains of a dead animal. A vulture swoops down hoping to share in the feast. But the vulture must be beware of the storks' big bills.

bee-eaters

ostriches

secretary bird

oxpecker

kind of bird. When a quelea flock lands on a field of grain the birds do enormous damage. Queleas weave nests of grass in trees. Each nest has an entrance hole in the side.

Insect-eating birds fly above the grasslands. The brilliantly coloured bee-eater is one of them. Bee-eaters are small, slim birds with pointed beaks. They are so called because they mainly eat bees and wasps; but they also eat locusts and other insects. The birds swoop down from a perch and catch their prey on the wing. Ox-peckers also eat insects. Oxpeckers are relatives of starlings. They sit on the back of a buffalo or rhinoceros and pluck out tiny creatures called ticks that burrow in its skin. This keeps the animal clean and healthy.

Ostriches are the world's largest birds. They live on insects and other small animals. Often, ostriches feed with herds of grazing mammals such as zebras. They eat creatures that the animals disturb. At the same time, they watch for danger. Ostriches are so tall that they can see an enemy far off. They cannot fly but run fast on long, strong legs.

Africa's open grasslands are hunting grounds for hawks, falcons, eagles and other birds of prey. One hawk hunts for lizards. Lanner falcons swoop on birds smaller than themselves. Ver-raux's eagle kills rock hyraxes, creatures about the size of rabbits.

Perhaps the bravest hunter is the secretary bird. This long-legged bird kills snakes by kicking them and beating them with its wings.

Vultures are big, black, ugly birds. They feed mainly on large animals that are already dead. The birds soar high in the sky looking out for dead or dying beasts. When a vulture finds food, others follow it. Soon they are all tearing at the corpse. Marabou storks sometimes share in the feast. Marabou storks and most vultures have bald heads. This means they can dig into a body without its blood messing up their feathers.

vulture

marabou storks

Birds in Woods and Fields

Many birds find food and shelter among trees. Others make their home in fields and moors. Some birds live partly in woods, partly in the open.

The different kinds of woodland bird find food in different places. Small birds, such as tits and warblers, flutter among leafy twigs in search of insects. Finches look for seeds in shrubs and flowering plants. Thrushes feed on berries, and hunt for worms on the ground. Woodpeckers patrol tree trunks and peck insects from the bark. Birds of prey, like owls and falcons, fly among the trees or overhead. Their sharp eyes watch for birds, mice and other small creatures.

Woodland birds find nesting places in trees or near them. Pigeons make their nests of twigs among the branches. Woodpeckers peck nesting holes in dead tree trunks. Finches and some warblers build their fragile homes among the leaves of shrubs growing under the trees. The woodcock makes her nest among dead leaves on the woodland floor.

long-eared owl

The nuthatch runs up and down trees hunting insects with its sharp bill.

nuthatch

treecreeper

The treecreeper walks along branches using its long, curved beak to dig insects out of the bark.

wood warbler

crossbill

The crossbill uses its scissor-shaped beak to tear open a pine cone. It eats the seeds inside.

Wood pigeons live in woods but often feed in fields. They are a pest to farmers as they eat crops.

acorn woodpecker

The acorn woodpecker pecks a hole in a tree, then stores an acorn in the hole. The bird eats the acorn later.

wood pigeon

woodcock

By day, woodcocks crouch on the woodland floor. At dusk, they dig for worms in soft earth. They often fly outside the wood to feed.

Birds belonging to open fields and moors have no trees where they can find food and safety from their enemies. Such birds as partridges and grouse peck grasses, or eat the shoots, berries and seeds of low-growing shrubby plants. Many birds that live and nest on the open ground have colours that match the plants or soil around them. This helps the birds to hide from enemies, especially when they are sitting on their nests.

Certain birds spend some time in woods and some time in the open. Starlings, for example, feed on worms and grubs found in meadows. But when evening comes, the birds fly into trees for the night. They are safer there. Many finches nest in woods but leave them in winter when seeds become scarce.

Pheasants live in fields and woods. The male's bright colours help it attract a mate. The female's dull-brown colour makes it easier for her to hide from enemies as she sits on her eggs. Pheasants nest on the ground. Often, the nest is outside a wood, in long grass at the edge of a field.

Ravens live in forests and on moors and mountains. They eat insects and kill rabbits. But they mainly eat dead animals, including sheep. Sometimes, they steal eggs from other birds. Ravens build a nest of sticks and earth. The nest can be perched in a tree or on a ledge high up a cliff.

The lapwing nests in fields and on moors. The nest is just a dip scraped in the ground and lined with stems. But the eggs and chicks are quite safe as their colours make them difficult to see.

raven

male pheasant

female pheasant

lapwing

lapwing chicks

77

Birds on Rivers and Marshes

Various birds live on rivers, lakes and marshes. Some swim out in the deep. Others wade through shallow water. Some fly over the surface. Others hide among reeds.

Ducks, geese and swans swim along the open water of lakes and rivers. They paddle with their feet. Swans and some ducks lower their heads to feed below the surface. Diving ducks and dippers plunge underwater to feed. Some ducks and geese come ashore to eat grass. Wading birds walk through shallow water round the edges of lakes and rivers. Using their sharp beaks they dig into the mud for food. Shy birds, like the water rail and bittern, hide among the reeds growing by marshes and lakes. Swallows and some other land birds often fly over water to catch flying insects.

swallow

A swallow swoops low over the water to catch flying insects.

Canada geese

cygnets

mute swan

Canada geese eat some water plants but they generally feed on grass on the shore.

A mute swan and her young ones, called cygnets. She is carrying two of the cygnets on her back. The mother does this when the young ones are too tired or too cold to swim.

kingfisher

The kingfisher has just dived into the river after a fish. Water drips from the bird's wings as it flies back to its perch on a riverside branch.

78

The long-legged heron stands still in shallow water waiting for fish. When it sees one, the bird makes a sudden lunge with its long, sharp beak.

heron

bittern

mallard duck

If danger is near, a bittern freezes with its beak pointing upwards. This makes it look like the surrounding reeds.

The water rail is a shy bird that hides among the reeds. At night, the water rail comes out to eat plants and insects.

water rail

A mallard drake (male duck) flies up from the water. Mallards live on rivers, lakes and ponds, even in city parks.

This coot keeps watch on her young family. Her nest is hidden among water plants where it is safe from land enemies such as rats and cats.

coot

young coots

Birds Beside the Sea

Many birds make their home by the sea.
Some feed out at sea, some along the shore.
Some nest on cliffs, some on beaches.

Different shore birds find food in different ways. Wading birds with long legs search the mud for shellfish. Oystercatchers and curlews are wading birds. Gannets dive into the water after fish. Some sea ducks dive to the seabed for shellfish.

curlew

common tern

The male tern presents a fish to the female.

little terns

Terns are sometimes called sea swallows. These graceful birds fly along the coast watching the sea for small fish. When the tern spots a fish, it dives down, seizes its prey from the water and then flies up again.

Terns nest in groups called colonies. A male attracts a female by flying over the colony with a fish in his beak, and calling. A female flies up in front of him. They land, and he offers her the fish. The birds mate and then scrape a nest in the sand. The female lays up to three eggs and both parents take turns at sitting on them. Soon after hatching, the chicks can leave the nest.

Several kinds of seagull live by the sea. Gulls are not clever fishermen, so they do not fly far from the shore. They search for dead fish and titbits washed upon the beach. Sometimes gulls fly inland to find food on rubbish tips and in plowed fields. In the picture, two herring gulls take fish from a box in a fishing port. Gulls nest on cliff tops and among sand dunes where they steal eggs and chicks from nesting terns.

herring gulls

gannet

young gannet

Gannets spend most of their lives out at sea. They glide and fly above the waves. Sometimes a gannet closes its wings and plunges down into the sea after a fish. In spring, gannets form huge colonies on the ledges of sea cliffs. Each female lays one bluish-white egg in a seaweed nest. Both parents take turns to sit on the egg which hatches in six weeks. The parents feed the young bird for nine weeks. Then it must look after itself.

oystercatchers

fulmar

puffin

Different shore birds have different kinds of wings. The fulmar has long, narrow wings designed for gliding. Like most seabirds, it has to be a powerful flyer because of the strong winds that blow over the oceans. The fulmar is a close relation of the albatross which has wings twice as long as a man. The fulmar and the albatross stay out at sea for months. They come ashore only to nest and raise their young.

Oystercatchers also fly strongly. They often utter shrill cries as they fly. On the other hand, puffins are not good flyers. They have short, stubby wings and spend a lot of time just floating on the sea.

Razorbills are expert at diving and at underwater swimming. They can plunge six metres (20 ft) deep into the sea and hold their breath under water for almost a minute. Meanwhile they hunt for fish. Razorbills swim quickly by pushing the water back with their wings. When a razorbill catches a fish, it stores the fish under its tongue. This stops the fish falling out when the razorbill opens its beak to catch another fish.

Razorbills sometimes float on the sea in huge flocks called rafts. They then take part in special courting displays. Some of the birds shake their heads from side to side. Others dive and rise from the water with their beaks pointing upwards.

cormorant drying its wings

cormorants

Cormorants have web feet, like other seabirds, but do not have waterproof feathers. After diving, they stretch out their wings to dry them.

razorbill

81

Birds in Towns and Cities

Although birds are generally afraid of people, some kinds choose to live in busy towns and cities.

Towns and cities make good homes for various birds. Many of them, like sparrows, feed on food scraps dropped by people. At night, starlings and others find warmth and shelter on buildings. Ducks and gulls swim in safety on park lakes. Pigeons often nest on window ledges.

This pied wagtail has central heating! It has built its nest on a building beside an outlet of warm air.

sparrows

pied wagtail

House sparrows are at home anywhere in cities.

house martin

The house martin's mud nest is glued to the wall under a ledge.

house martin

City pigeons often build their nests on high window ledges.

At dusk, starlings line up on a ledge and prepare to roost.

starlings

black-headed gull

tufted duck

mandarin duck

moorhen

A male pigeon struts about as he shows off to a female. This is a common sight in the streets of cities and towns.

female pigeon

male pigeon

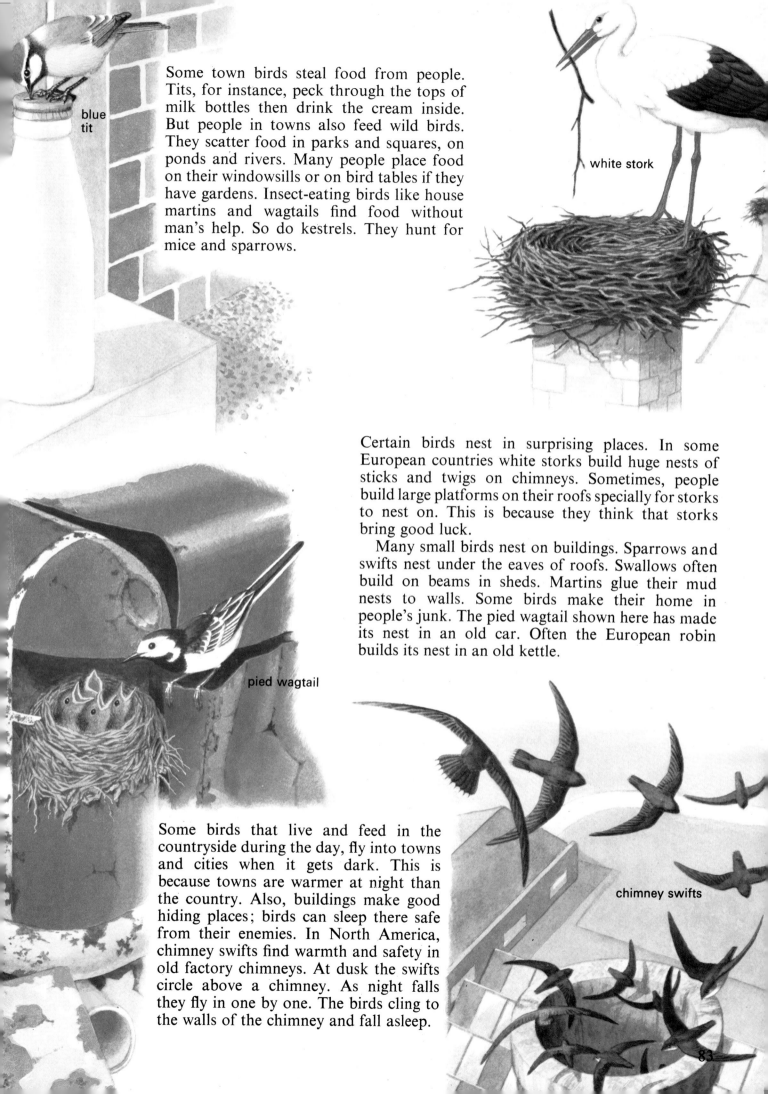

blue tit

Some town birds steal food from people. Tits, for instance, peck through the tops of milk bottles then drink the cream inside. But people in towns also feed wild birds. They scatter food in parks and squares, on ponds and rivers. Many people place food on their windowsills or on bird tables if they have gardens. Insect-eating birds like house martins and wagtails find food without man's help. So do kestrels. They hunt for mice and sparrows.

white stork

Certain birds nest in surprising places. In some European countries white storks build huge nests of sticks and twigs on chimneys. Sometimes, people build large platforms on their roofs specially for storks to nest on. This is because they think that storks bring good luck.

Many small birds nest on buildings. Sparrows and swifts nest under the eaves of roofs. Swallows often build on beams in sheds. Martins glue their mud nests to walls. Some birds make their home in people's junk. The pied wagtail shown here has made its nest in an old car. Often the European robin builds its nest in an old kettle.

pied wagtail

Some birds that live and feed in the countryside during the day, fly into towns and cities when it gets dark. This is because towns are warmer at night than the country. Also, buildings make good hiding places; birds can sleep there safe from their enemies. In North America, chimney swifts find warmth and safety in old factory chimneys. At dusk the swifts circle above a chimney. As night falls they fly in one by one. The birds cling to the walls of the chimney and fall asleep.

chimney swifts

83

In a Rock Pool

Some seashore plants and animals live in rock pools. Here they find shelter from storms, hot sun and cold winds.

When the tide goes out from a rocky shore, some of the seawater stays behind in pools. Rock pools are exciting places to explore. At a quick glance, they may seem still and empty. But really, their waters are full of life. The plants and animals shown on these pages make their home in rock pools.

Some of the plants and animals found in rock pools are very difficult to see. Shrimps hide in the sandy floor with just their eyes showing. Small, slim fish such as butterfish and gobies slide into narrow crevices between the rocks. Crabs crouch under overhanging stones and boulders. But most of these animals come out from time to time to feed on even smaller sea plants and creatures.

Other kinds of rock-pool life are easier to see. Seaweeds wave under the water or lie limply on rocks exposed by the falling tide. Their special root-like holdfasts keep seaweeds fixed to the rocks, even when wild storm waves try to tear them off.

This lesser black-backed gull has just caught a small crab in a rock pool. Gulls often stand by rock pools at low tide. There they can always make a meal out of shellfish and other sea animals. Some hungry gulls even eat seaweed.

mussels

dog whelk

topshells

sea anemone

prawn

edible sea urchin

crab

starfish

butterfish

84

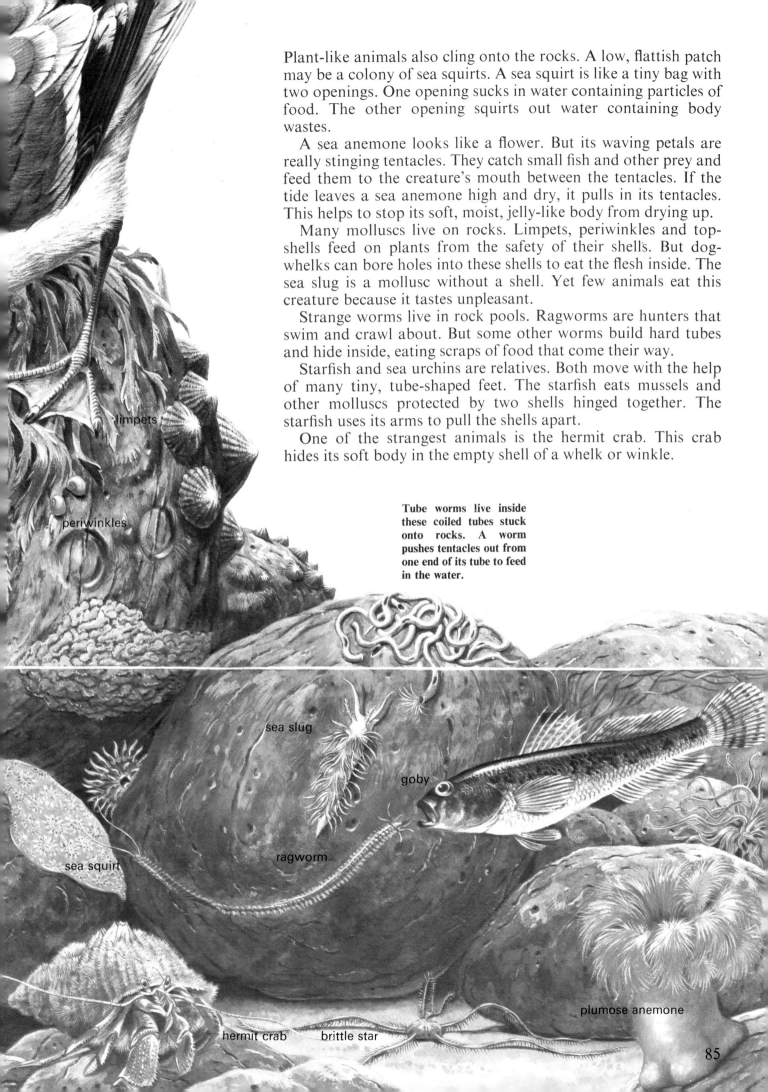

Plant-like animals also cling onto the rocks. A low, flattish patch may be a colony of sea squirts. A sea squirt is like a tiny bag with two openings. One opening sucks in water containing particles of food. The other opening squirts out water containing body wastes.

A sea anemone looks like a flower. But its waving petals are really stinging tentacles. They catch small fish and other prey and feed them to the creature's mouth between the tentacles. If the tide leaves a sea anemone high and dry, it pulls in its tentacles. This helps to stop its soft, moist, jelly-like body from drying up.

Many molluscs live on rocks. Limpets, periwinkles and top-shells feed on plants from the safety of their shells. But dog-whelks can bore holes into these shells to eat the flesh inside. The sea slug is a mollusc without a shell. Yet few animals eat this creature because it tastes unpleasant.

Strange worms live in rock pools. Ragworms are hunters that swim and crawl about. But some other worms build hard tubes and hide inside, eating scraps of food that come their way.

Starfish and sea urchins are relatives. Both move with the help of many tiny, tube-shaped feet. The starfish eats mussels and other molluscs protected by two shells hinged together. The starfish uses its arms to pull the shells apart.

One of the strangest animals is the hermit crab. This crab hides its soft body in the empty shell of a whelk or winkle.

Tube worms live inside these coiled tubes stuck onto rocks. A worm pushes tentacles out from one end of its tube to feed in the water.

limpets

periwinkles

sea slug

goby

sea squirt

ragworm

hermit crab

brittle star

plumose anemone

Safe in the Sand

Along a sandy shore, there are no pools for creatures to shelter in or rocks for them to cling to. To find moisture and safety, sea animals must burrow into the sand.

At low tide a sandy beach seems deserted. Yet millions of shore creatures are hiding under the surface. These animals can only breathe in water and must therefore keep themselves moist. Buried in the sand they stay comfortably cool and damp even if the surface bakes in the sun or is chilled by a frost. When the tide rises, many of these burrowing creatures crawl out of the sand. Others push up tubes or tentacles to catch particles of food washed in with the waves. But if a fierce storm is raging and angry seas batter the beach, the tiny animals stay safely under the sand's surface.

The pictures on these pages show some of the creatures found on a sandy shore. Each one has its own way of living and moving about among the sand grains.

A shrimp burrows into the sand by kicking with its feet. It also uses its gills to squirt sand away. When the shrimp has made a dip in the beach, it snuggles down into it. Then, the animal spreads sand over its back with its feelers.

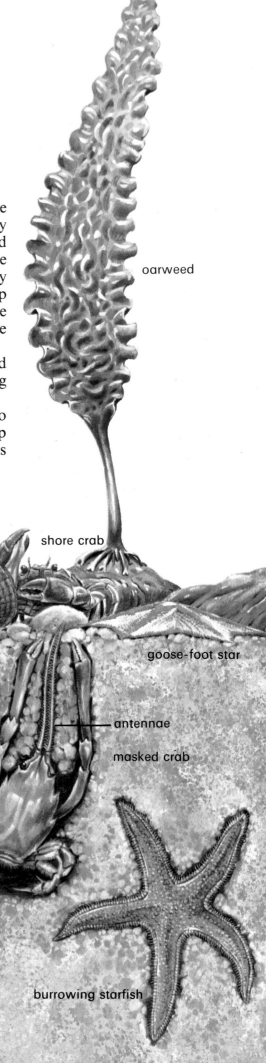

oarweed

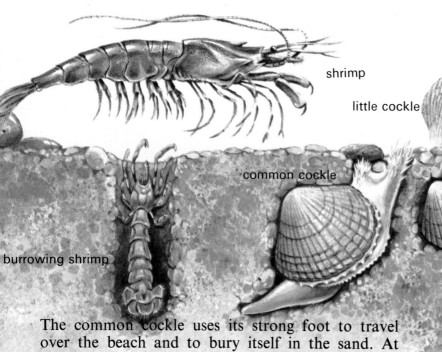

shrimp

shore crab

little cockle

common cockle

goose-foot star

burrowing shrimp

antennae

masked crab

The common cockle uses its strong foot to travel over the beach and to bury itself in the sand. At high tide, cockles lie just under the surface. They push two tubes up through the sand into the water. One tube sucks in water and food particles. The other squirts out water and wastes.

The masked crab burrows backwards into the sand. The two long feelers or antennae on its head form a tube that reaches the top of the sand. Water runs down the tube so that the crab can breathe even while it is buried in the sand.

The shore crab usually crouches under stones but it can use its legs to tunnel into sand.

burrowing starfish

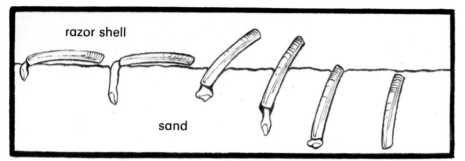

razor shell

sand

The amphitrite is a long worm that lives in a burrow. A mass of tentacles sprouts from its head. When the tide is in, these wave about. Bits of food stick to them. Hairs on the tentacles move food to the amphitrite's mouth.

The purple heart urchin and the sea potato are sea urchins. The sea potato's long tube feet help it to feed and breathe. But it burrows with its spines.

Tellins are molluscs that stay buried while they suck in food.

The goose-foot star is so named because it is shaped like the webbed foot of a goose. This starfish lies on the sand and eats shrimps and molluscs. But the burrowing starfish uses its tube feet to dig deeply into the sand. This starfish eats cockles and other molluscs. It takes a whole cockle into its stomach. If the cockle slightly opens its hinged shell, juices from the starfish's stomach kill and digest the cockle.

The razor shell burrows by stretching out its strong foot and pulling the shell down after it; the picture shows how this happens. No other mollusc digs so fast.

The lugworm lives in a U-shaped burrow. It swallows mud containing bits of plants and dead animals. Used mud squirts from its tail and coils up on the sand.

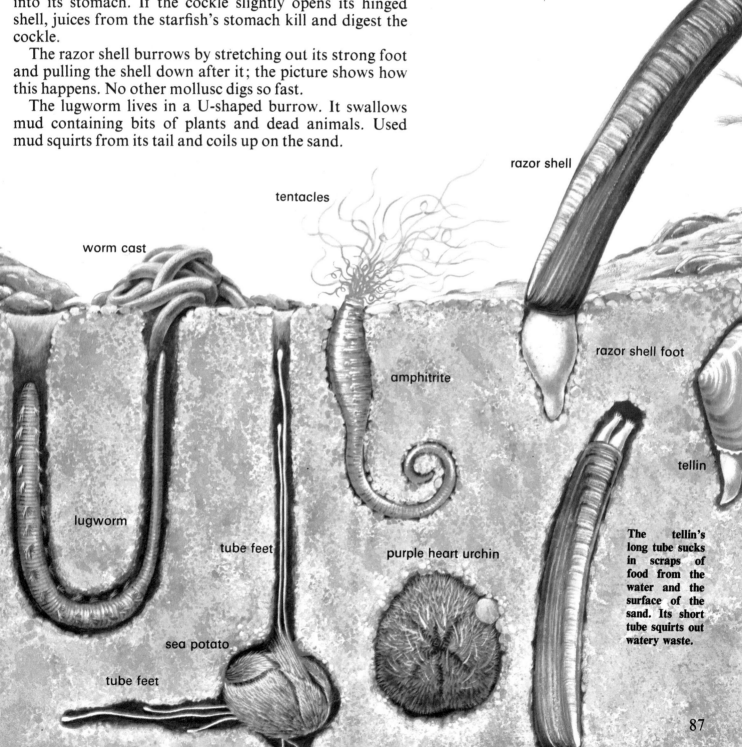

razor shell

tentacles

worm cast

razor shell foot

amphitrite

lugworm

tube feet

tellin

tube feet

sea potato

purple heart urchin

The tellin's long tube sucks in scraps of food from the water and the surface of the sand. Its short tube squirts out watery waste.

87

The Crab Family

Crabs, lobsters, prawns, shrimps and barnacles are all crustaceans. Most crustaceans have a hard crusty shell and several pairs of jointed legs.

Many crustaceans are sea creatures. Some live on the shore, others in deeper water. Shrimps like to burrow in a sandy floor, but most of the crustaceans shown on these pages live along rocky shores or in deep water nearby.

Crabs live in the sea and along the shore in rock pools or under stones and seaweed. The shore crab is small but it has strong claws for seizing prey and pinching enemies. If disturbed, this crab quickly scuttles off to find another hiding place. A crab moults several times as it grows. Each time it sheds its shell, the crab hides its soft body until it grows a new hard coat.

The edible crab is much larger than the shore crab and has bigger, thicker claws. In summer it lives near the shore. In winter, it moves out to warmer, deeper water. Many people like to eat edible crabs.

Hermit crabs never grow a hard shell. Instead, they live in the empty shells of sea snails. When a hermit crab is young and small it lives in a winkle shell. A full-grown hermit crab chooses a whelk shell for its home. Sometimes the crab carries a sea anemone on the shell. The sea anemone

bladder wrack

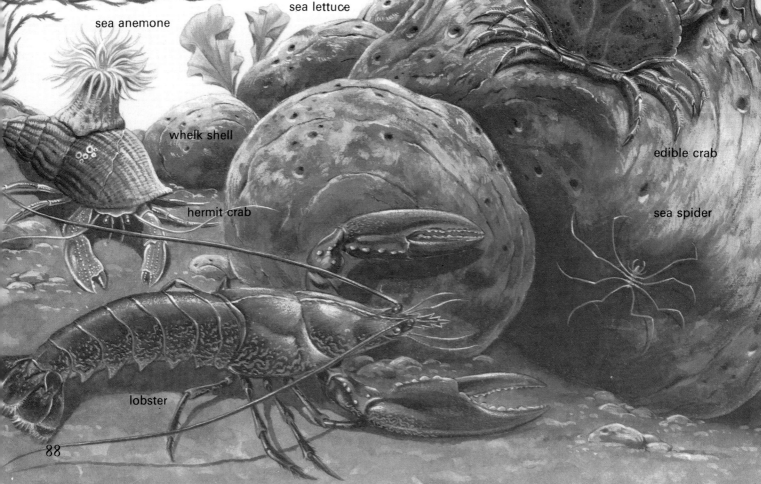

serrated wrack

sea lettuce

sea anemone

whelk shell

hermit crab

edible crab

sea spider

lobster

88

eats food scraps dropped by the crab. In return, the sea anemone's stinging cells guard the crab against enemies such as squids and octopuses.

Shrimps and prawns look much like each other. But shrimps are smaller and flatter than prawns. Also shrimps have much larger front legs that end in hooks. Prawns, but not shrimps, have a long, sharp spear jutting forward between the eyes. Shrimps, prawns and many other crustaceans are covered with hard, shell-like plates rather like knights in suits of armour. As crustaceans grow, they shed their old suits and develop new ones underneath.

The lobster lives off rocky shores. It swims backwards by bending and stretching its body to push water forward with its flat tail. The lobster has five pairs of legs. Four pairs are for walking. The other pair, at the front, ends in large, strong pincers. The lobster uses these to crush its prey. A lobster can even break the hard shell of a crab. Lobsters are blue when alive, but pink when cooked.

Barnacles start life as tiny creatures that swim about. In time they settle down and change their shape. Acorn barnacles look like little pointed hats stuck onto rocks. Each hat is a shell made of hard plates. When the tide is in, the plates open and out pop six pairs of feathery legs. These make grabbing movements that trap food particles. When the tide goes out, the acorn barnacle pulls its legs inside and shuts its shell. There is enough water inside to keep the animal alive until the tide comes in again. Acorn barnacles live for five years or more.

The goose barnacle's body is protected by hard, shell plates that are very like the plates round the acorn barnacle. But unlike the acorn barnacle, the goose barnacle does not crouch low on rocks. Its body hangs from the end of a long rubbery stalk. The other end of the stalk is fixed to a floating object such as a piece of drift-wood, a bottle or even the bottom of a ship.

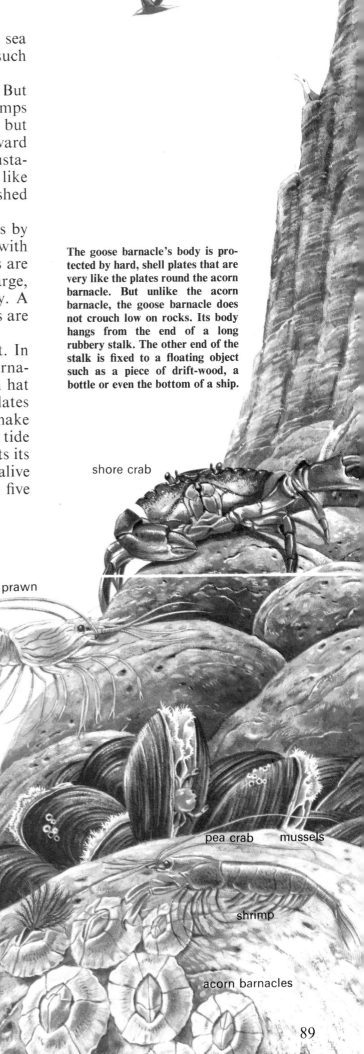

shore crab

prawn

goose barnacles

The small, soft-bodied pea crab hides in a mussel shell. The animal strains tiny bits of food from the water.
The sea spider has jointed legs like the crustaceans. But this creature is related to the true spiders. Sea spiders feed on plant-like animals including sea anemones.

serrated wrack

pea crab   mussels

shrimp

acorn barnacles

89

Swimming Inshore

Only fishes that do not mind rough water can live close to the shore. Some escape waves, and enemies, by hiding in sand. Or between rocks. Many shore fishes have shapes that help them to hide.

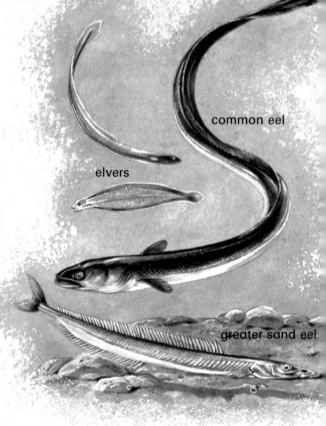

common eel

elvers

greater sand eel

Two of the fishes seen here, the greater sand eel and the common eel, look rather like snakes. Both of them are sometimes found along the shore. The sand eel uses its long lower jaw as a spade to dig itself down into the sand. The common eel's real home is in lakes and rivers. But full-grown eels swim down to the sea to breed. They travel far across the Atlantic Ocean to spawn. Then they die. The eggs they have laid hatch into young eels. These tiny creatures swim and drift with ocean currents back to the shores their parents came from. Some go to America. Some to Europe. The young eels, now called elvers, then swim up rivers to make their home inland.

shanny

sand goby

The shanny is often seen darting across a shallow rock pool. This small fish is also called the common blenny. It has a large head, strong jaws and can bite off barnacles stuck to rocks. Yet the shanny is quite shy and hides under stones. It makes itself a shelter by wriggling down into the sand or gravel next to a stone. In summer the female shanny lays yellow eggs in crevices or under rocks. The male then guards the eggs. He fans his tail to keep water moving over them. This helps them to get more air from the water.

The male sand goby also guards the female's eggs. She lays these in a place that he has chosen.

Flounders and plaice are flatfishes. Like other fish, a young flatfish has one eye on each side of its head. But, as it grows, one eye slowly moves round the head to join the other. Then the flatfish lies on the sea bottom with both eyes facing up. The fish's colouring matches the sand and stones, so its enemies find it difficult to see. Flatfish cannot hang in water like most fish. If they stop swimming they sink.

flounder

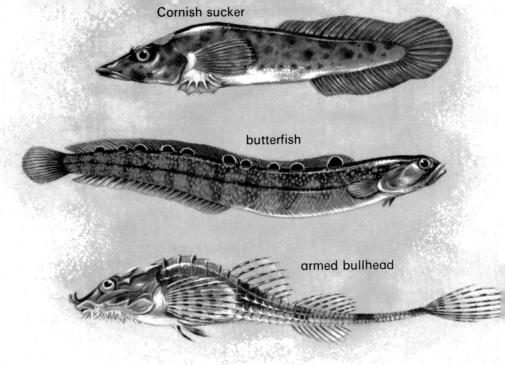
Cornish sucker

butterfish

armed bullhead

The Cornish sucker has some fins shaped to form a sucker. This helps the fish cling to rocks as rough seas surge above it.

The butterfish has a slimy body as slippery to hold as butter. When the tide drops, this eel-like fish hides under damp rocks, stones or gravel.

The armed bullhead has an odd turned-up nose armed with spines. Bony plates protect its body, rather like a suit of armour. This fish likes swimming in river mouths and harbours.

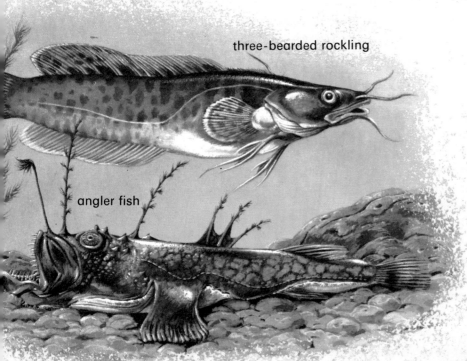
three-bearded rockling

angler fish

The three-bearded rockling has three fleshy whiskers round its mouth. One grows from the chin, the others from the snout. All act as feelers. They help the rockling feel for food in the dark nooks and crannies of rock pools. Rockling belong to the same family as cod.

The angler fish has a fishing rod growing from its head. This rod is a long stalk with a tip looking like a tasty worm. When the angler fish waves its rod slowly in the water, small fishes swim up to eat the 'worm'. Then the fish seizes them. Angler fishes can grow as long as a man. Although they are sea fish, they sometimes swim some distance up the mouths of rivers.

Seahorses and their relatives, the pipefishes, do not resemble any other kind of fish.

The seahorse has a head shaped rather like a horse's head. The creature swims along, upright, by waggling the fin on its back. A seahorse clings to seaweed with its tail just as some monkeys use their tails to cling to branches. Males have stomach pouches where they store eggs laid by females. Most seahorses live in warm seas.

Pipefishes are long, thin, pipe-shaped animals with small mouths.

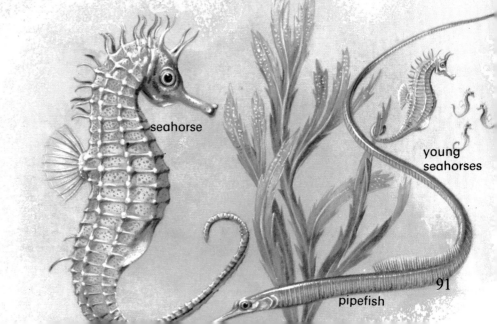
seahorse

young seahorses

pipefish

Life in Warm Seas

In warm parts of the world, the shallow water round the coasts are filled with life. Many creatures make their home along these sunny shores.

Big sea turtles live in warm, shallow seas. These reptiles are strong swimmers. Their large, flat front legs work like oars pushing the water backward. Their short, broad hind legs act as rudders. Turtles must swim to the surface to breathe. But they find food in the water. Green turtles eat underwater plants. Most other turtles eat animals. Hawksbill turtles catch stinging jellyfish. Ridley and loggerhead turtles hunt crabs. Leatherbacks, the largest turtles of all, eat squid and fishes.

Turtles mate at sea, but the females come ashore to lay their eggs. Above high water mark, they dig deep holes in the sand with their hind flippers. Each female lays 50 to 200 white eggs as big as billiard balls. After laying, the mother hides the eggs by filling the hole with sand. Then she flops back down the beach into the sea.

Warmed by the sun's heat, the eggs hatch about two months later. The baby turtles quickly scramble down the shore. But crabs, dogs, gulls and other enemies catch most of them. Fewer than one in a hundred survive to swim away.

Above: Four pictures show how a baby turtle starts life. In the top picture the baby breaks out of its egg. Then the young animal pushes up through the sand to reach the open air. The next picture shows several baby turtles climbing out onto the beach. The last picture is of the shore seen from high in the air. Gulls and frigate birds are circling in the sky, waiting to pounce on the little turtles as they scuttle down to the sea.

turtle

92

Huge numbers of fiddler crabs are found along warm, muddy shores. The crabs live in holes in the mud. They eat particles of food from mud scooped up into their mouths. Fiddler crabs can remain in air a long time before they have to go into the water to wet their gills again. Each crab stays near its burrow and keeps other crabs away, threatening them with its claws. The male fiddler crab has one claw much larger than the other. Often, the large claw is a different colour from the rest of the body. This claw looks like a fiddler's arm, holding his bow. Although their large claws seem to be powerful weapons, the males seldom use them for defence. Instead, they wave them in the air to attract female fiddler crabs.

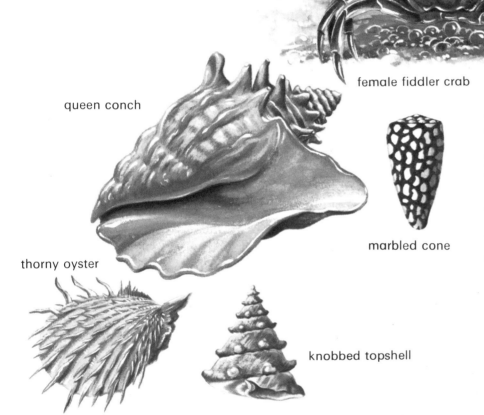

male fiddler crab

female fiddler crab

queen conch

marbled cone

thorny oyster

knobbed topshell

Many of the molluscs belonging to warm seas have beautiful shells. They often have bright colours and patterns, and some grow very large. The queen conch from the Caribbean Sea has a shell as long as a man's foot. The thorny oyster shell is protected by long spines that make it look rather like a chrysanthemum flower. People sometimes call it a chrysanthemum shell. The marbled cone lives in the sand along coral reefs. Cone shells kill sea worms and small fish with a poisonous bite. Some can even kill a man. The knobbed topshell is one of more than 1000 different kinds of topshell.

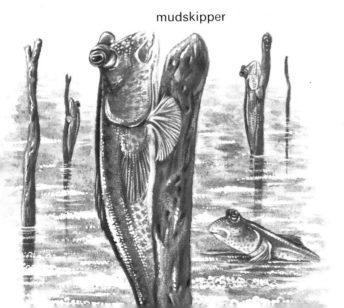

mudskipper

Mudskippers are small fish with large heads. They live around mudflats and mangrove swamps by the edge of the sea. When the tide goes out, mudskippers stay on the muddy shore. They can breathe in air for some time and use their fins like legs to crawl and hop over the mud. Some mudskippers have special sucker-like fins for hauling themselves up spiky mangrove roots. A mudskipper's eyes stick out on little stalks. This helps the fish to see in all directions. If danger threatens, the fish quickly jumps or skips away.

Most mudskippers hunt sandhoppers and tiny crabs. But there are some mudskippers that chew seaweed and others that feed on tiny plants in the mud.

Life in Cold Seas

In the far north and far south, the coasts are cold and bare. Yet many animals live there. They have special ways of surviving ice and snow.

The shores and waters of the Arctic and Antarctic Oceans are frozen for much of the year. Most small shore creatures, like crabs, cannot survive such cold. So, there are very few of them in polar regions. Yet there are plenty of fish in the icy seas. These fish provide food for many different birds and mammals.

Strange birds that cannot fly live in the far south on the shores of the Antarctic Ocean. These birds are penguins. Penguins swim and dive for fish and other sea foods. Their feathers and thick body fat help to keep out the cold. Penguins rest and breed on shore. The largest penguin, the emperor, has unusual breeding habits. In autumn, emperor penguins come ashore and find a mate. Later, each female lays just one egg on the bare ice. She then returns to the water to fish for food. Her mate places the egg on his feet and tucks it under a warm flap of skin. He stands like this on the ice without a meal for sixty cold, dark, winter days keeping the egg warm. When it hatches, the female feeds the chick on fish brought up from her stomach.

The emperor penguin chick keeps safe and warm nestling between its father's feet.

emperor penguin

penguin chicks

polar bear

polar bear cubs

ptarmigan

Below: A ptarmigan in its white winter feathers. This land bird lives along the Arctic shores all round the year. In winter it manages to find food, such as shoots, leaves and berries, hidden in the snow. Sea and shore birds, however, only come to the Arctic to breed. They come in summer when they can find food in the sea or in inland waters. But when the short summer ends, the waters containing food freeze over. Then millions of geese, ducks, divers, terns and wading birds fly south to warmer lands and waters.

The polar bear is master of the frozen lands round the North Pole. These big white bears kill seals as they surface to breathe through holes in the sea ice. The polar bear swims strongly in icy water without becoming cold. It often swims far out across the sea.

Arctic foxes also travel out to sea, but only over ice. These foxes sometimes follow polar bears to feed on scraps left over from their meals.

Many kinds of seal swim in cold seas. A thick layer of fat under the skin keeps them warm. Their legs are shaped like flippers to help them swim easily through the water. Seals move awkwardly on land. True seals pull themselves along like giant maggots. Sea lions and walruses can just manage a clumsy gallop.

Walruses live in groups. They use their long tusks to rake shellfish from the seabed. Then, as they eat, their bristling moustaches strain the mud from their food. Walruses can eat hundreds of clams in one meal. A baby walrus mainly feeds on its mother's milk until it is two. Its tusks are then long enough for it to dig up shellfish. When walruses lie on warm beaches, they glow a rosy red. The colour comes from blood filling tiny tubes in the skin. Heat escapes from the blood and this helps to keep walruses comfortably cool.

Whales also feed in cold oceans. These giant mammals never come ashore at all. They can only use their flippers like fishes' fins.

A female snowy owl seizes a lemming. Both creatures often live near the sea but feed on land.

female snowy owl

lemming

Below: The musk oxen live on land, but the other Arctic animals in the picture depend in some way on the sea. The walruses eat underwater clams. The humpback whale swims in northern waters to feed on tiny shrimp-like creatures called krill. The Arctic foxes are scavenging for food along the shore.

whale

arctic foxes

musk oxen

musk ox calf

walrus

walrus pup

95

At Home in Deep Water

The great oceans are alive with fish. Some live in the sunny surface waters; some swim through the dark middle depths; others belong to the blackness of the ocean floor.

flying fish

Portuguese man-o'-war

basking shark

jellyfish

tuna

seal

seahorse

swordfish

anchovies

animal plankton

crab

plant plankton

squid

giant squid

giant swallower

oarfish

prawn

hatchet fish

glass sponges

1000m
(3280ft)

viper fish

squid

prawn

grenadier

3000m
(10,000ft)

gulper eel

angler fish

tripod fish

crinoids

lamp shells

sea cucumber

96

glass sponges

brittle-star

gannet

sea anemones

sea snail

seaweed

This picture shows some of the plants and animals that live in the three main levels of the ocean.

In the surface layer tiny plankton plants and animals form food for creatures like anchovies, squid, basking sharks and sea-horses. Near the shore, a sea snail grazes on seaweeds and other plants growing on the rocks. A crab has also made its home on the rocks. Flying fish, anchovies and squid are all prey for hunters like the swordfish, tuna, seal and gannet. Some fish are eaten by the plant-like sea anemone and by the jellyfish and the Portuguese man-o'-war. Both drift with the ocean currents and kill by using their stinging tentacles.

The middle layer of the ocean contains far fewer animals than the surface layer. The picture shows six very different creatures. The oarfish is long and thin with a top fin like a ribbon, but no tail. Oarfish can grow five times longer than a man. Hatchet fish are shorter than a man's hand and some are only as thick as a coin. The giant swallower is small but has huge jaws and a colossal stomach. It can eat fish twice its own size. The giant squid has long tentacles for grasping prey.

The bottom layer of the ocean is the home of small fish with huge jaws. The viper fish and gulper eel are examples. The angler hunts other fish using its lighted lure. The tripod fish has long fins that let it rest on the soft seabed without sinking in. The sea cucumber, crinoids and other bottom dwellers get food scraps from mud or water.

All creatures in the ocean depend on plants for food. Some animals eat the plants themselves. Others eat animals that eat plants. But plants grow only in the surface layer of the sea. Only here do plants find enough sunlight to supply them with the energy they need for making food. Most of these plants are tiny organisms that drift with the ocean currents. These drifting organisms are called plankton. There are millions of plankton plants. Some are little spiny organisms. The spines help the plants to soak up sunshine and to stay afloat. Other plankton plants swim weakly by lashing the water with tiny whips.

Plant plankton is food for many of the tiny drifting creatures that make up animal plankton. These microscopic plant-eaters include the copepods. Copepods are related to crabs. They use hairs on their legs to trap the tiny plants.

Copepods and other small plankton animals provide food for fish that swim near the surface. Most of these fish are small. A few are enormous. Basking sharks and whale sharks eat plankton, yet these are the largest fish in the world. At night, deep-sea squid and prawns swim up to join in the feast. But danger is always near. Large fish prey upon small fish and other plankton eaters. Swordfish and tuna are among the hunters. Birds and mammals, such as gannets and seals, also take their share.

Ocean waters get darker as they get deeper. Plants cannot grow in the dark. So there are fewer plants at great depths than on the surface. This also means that fewer animals live deep down as there is not so much for them to eat. All of them depend on food that comes from above. Some are scavengers: they feed on the remains of dead plants and animals that sink down from the surface. But most eat the shrimps, squid and other deep-sea creatures that swim up at night to find food and then swim down again.

Many of the animals that live far below the surface seem strange. This is because they are suited to life in the middle or bottom layers of the ocean. At depths of 250 metres (820 ft) there are fishes with rows of lights and silvery sides like mirrors. Deeper down live fishes with lights that flash and wink as the creatures move flaps of skin. How these deep-sea fishes use their lights is still largely a mystery. Probably, the lights act as signals to attract mates. And perhaps, if danger is near, the fish switches off its light and becomes invisible.

Their eating habits are not such a mystery. Most fish living at the bottom of the ocean have huge stomachs. This means that when a meal does come their way—which is not very often—they can eat a great deal. Below 1000 metres (3280 ft) food is so scarce that fish must not waste energy swimming after it. Instead, creatures like the deep-sea angler fish have brightly shining baits that attract their prey to them.

Life is thinly spread across the bottom layer of the ocean. At these great depths most animals are scavengers. Sea cucumbers, lamp shells and other bottom dwellers suck in nourishment from animal remains or droppings. These remains drift down to the ocean floor from the surface waters far above.

Giants of the Oceans

Most sea animals are small, but some are huge. Earth's biggest fish are certain sharks and rays. But the giants of the oceans are not fish at all. They are mammals. Mammals must surface to breathe air. Sea mammals include whales and seals.

A walrus can weigh as much as eighteen men. The fur seal is a smaller animal.

walrus

bottlenose dolphin

The bottlenose dolphin has a snout like the neck of a wine bottle. It swims in schools of a hundred or more.

Dolphins are a group of small whales. They love acrobatics.

The swordfish's upper jaw ends in a long, sharp sword. The animal uses this sword to stun smaller fish so that it can then kill and eat them.

common dolphin

swordfish

sperm whale

Electric rays give electric shocks that can stun or kill an enemy. They are smaller, but more dangerous, than giant rays. Giant rays grow 6 metres (19½ ft) across.

Above: The sperm whale has teeth. It eats fish and squid. Left: The dugong feeds on water plants near the coast.

electric ray

dugong

96

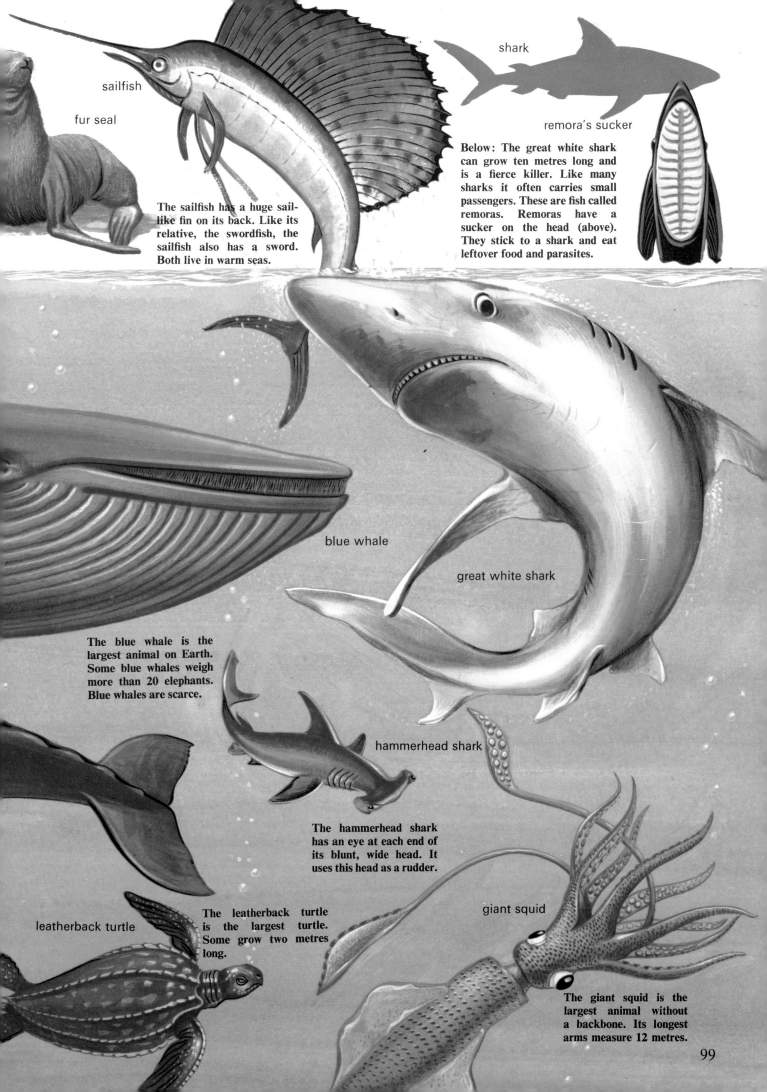

sailfish

fur seal

The sailfish has a huge sail-like fin on its back. Like its relative, the swordfish, the sailfish also has a sword. Both live in warm seas.

shark

remora's sucker

Below: The great white shark can grow ten metres long and is a fierce killer. Like many sharks it often carries small passengers. These are fish called remoras. Remoras have a sucker on the head (above). They stick to a shark and eat leftover food and parasites.

blue whale

great white shark

The blue whale is the largest animal on Earth. Some blue whales weigh more than 20 elephants. Blue whales are scarce.

hammerhead shark

The hammerhead shark has an eye at each end of its blunt, wide head. It uses this head as a rudder.

leatherback turtle

The leatherback turtle is the largest turtle. Some grow two metres long.

giant squid

The giant squid is the largest animal without a backbone. Its longest arms measure 12 metres.

tiger

serval

The serval is an African cat about a metre long. It hunts guinea fowl and can leap two metres into the air to catch a bird.

Left: A tiger prowls through an Asian jungle. Each tiger kills more than forty deer, cattle or other animals every year. This means that tigers need big hunting grounds. But people have taken land for farms. They have also killed tigers as pests or for sport. So now, tigers are scarce. Once they lived in many parts of Asia. Today, there are few outside India.

jaguar

Big Cats

The cats people keep as pets belong to the same animal family as lions, tigers and jaguars. Pet cats are tame. All other cats are wild.

Cats are furry, flesh-eating mammals. They all have sleek, agile bodies, sharp teeth and strong claws. They use their teeth and claws to catch and kill animals.

The wild cat of western Europe looks very like a pet tabby cat. But most wild cats are much larger. They are known as the big cats and include lions and tigers. A full-grown lion or tiger can weigh twice as much as a heavyweight boxer. Yet, despite their size, lions and tigers are often difficult to see. The tiger's striped coat hides it among long grass. The lion's tawny fur blends with the dusty plains where it lives. Both these big cats are powerful hunters. They creep up on grazing mammals, such as antelopes, then make a sudden spring.

Long ago, there was an even larger hunter: the sabre-toothed cat. In its upper jaw, this huge, fierce creature had two long teeth shaped like sabres or curved swords.

Today, Africa's lions and Asia's tigers are the giants of the cat family. The largest American cat is the jaguar. The jaguar is as long as a man; longer if the tail is included. Jaguars live in the warm forests of South and Central America. They are agile climbers. Sometimes they hide in trees, waiting to catch monkeys or sloths. On the ground, they kill animals such as deer, cattle and wild pigs. Jaguars are also strong swimmers and catch fish and alligators. Like most cats jaguars are night creatures.

sabre-toothed cat

This leopard has pulled an antelope up a tree. There, the leopard's prey is safe from scavengers like jackals. A leopard chased by a lion can haul a dead antelope to the highest branches. On the ground, a cheetah has just caught and struck another antelope.

leopard

The leopard, found in Africa and Asia, looks very like a jaguar. Both animals have spotted coats that camouflage them in the patches of light and shade under trees. But the leopard is smaller and some Asian ones are almost entirely black. These are called panthers. As with other big cats, the leopard kills by suddenly springing on its prey.

Another big cat, the cheetah, also belongs to Africa and Asia. Over short distances, the cheetah is the fastest animal on Earth. Across the open plains this hunter runs down antelopes and kills them by springing on their backs. The cheetah's slim body and long, powerful legs help produce its great speed.

The cheetah is pale brown with black spots. Its claws are different from the claws of all other cats: they are blunt and cannot be drawn in.

cheetah

A lion with his cubs. Lionesses usually look after the young animals but, as they also do most of the hunting, the cubs are sometimes left with their father.

The bone and muscle of a cat's toe, showing how the claw can be pulled in.

lion cubs

101

Bears

Bears are large mammals with thick fur, strong jaws, powerful limbs and long claws. They eat many foods, but love sweet things.

Below: A polar bear breaks the ice into splinters as it strikes at a seal. The seal is swimming up to take in air at its breathing hole. Polar bears crouch in wait beside these breathing holes. When the seals reach the hole, the bears swipe swiftly with a paw. A polar bear is strong enough to lift a full-grown seal out of the water.

polar bear

seal

Although bears look cuddly, they can be fierce when they are angry or protecting their cubs. Most bears have a varied diet: they eat foods such as fish, roots, berries and small animals. Some raid bees' nests for honey. In general, bears are not very agile hunters. They walk flat-footed and cannot leap upon prey. Also, their sight is poor but their sense of smell is good.

There are several kinds of bear alive today. The bears of old fairy tales were European brown bears. Only a few of these are left. They live in lonely mountains and forests. American brown bears are much more common. There are two types: grizzly bears and Kodiak bears. Grizzly bears are found in the Rocky Mountains of North America. Kodiak bears belong to Alaska. Kodiak bears are the largest of all bears; they are also the largest flesh-eating animals that live on land. A big Kodiak bear stands almost three metres (ten feet) tall and weighs about two-thirds of a ton.

Polar bears are almost as large. These big bears live in the far north. In the snow and ice their white coat hides them from the seals and seabirds they hunt.

Below: A polar bear swims by paddling with its strong front legs. Its hind legs trail behind. If the water is rough, the bear ducks its head and only lifts it to breathe. A thick layer of lightweight fat underneath the skin helps to keep the animal afloat and warm.

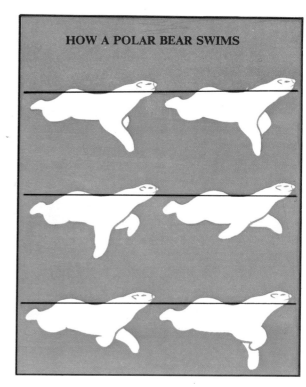

HOW A POLAR BEAR SWIMS

Polar bears are strong swimmers. They can also run fast enough to catch reindeer.

American black bears are much smaller then brown bears and polar bears. They can run quite fast and climb trees well. Another kind of black bear lives in the Himalaya Mountains of Asia. Asia is also the home of the sloth bear and sun bear. The sloth bear, from India, has black fur and a white mark on the chest. It loves eating insects and honey. The sun bear, just one metre (three feet) long, is the smallest bear of all.

The spectacled bear is the only bear in South America. The pale rings round its eyes look like spectacles.

This black bear cub clings to a tree trunk. Black bears climb well. If danger is near, the mother bear sends her cubs up a tree. They are safe there until the danger passes.

spectacled bear

grizzly bear

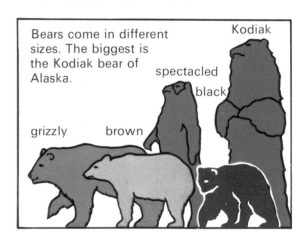

Bears come in different sizes. The biggest is the Kodiak bear of Alaska.

Kodiak

spectacled

black

grizzly   brown

A female brown bear has her cubs in winter while she is sheltering in a cave or other den. For weeks, she feeds the cubs on her milk. Then, in spring, the mother brings them out and teaches them to find food for themselves. Sometimes a brown bear family includes half-grown bears born two or three years before the youngest cubs.

brown bear family

103

Elephants

Elephants live in Africa and Asia. They are the largest land animals on Earth.

Elephants are giants: many bull elephants stand twice as high as a man and weigh as much as seven family cars. They have bigger ears, thicker legs, a longer nose and longer teeth than any other animal. Their wrinkled skin appears almost bald and is as thick as the heel of a man's shoe.

All these unique features make the elephant seem a rather strange creature. But each part of its enormous body serves a useful purpose.

female elephant

The elephant's long nose, or trunk, acts as a hand. The animal uses it to lift food and water to its mouth. The trunk is sensitive enough to pick up a tiny grape, yet strong enough to lift a heavy log.

A full-grown elephant has two kinds of teeth. Inside its mouth are big, flat-topped teeth used for chewing food. Outside its mouth, on each side of the trunk, are two long curved teeth, or tusks. The elephant uses its tusks for lifting loads and fighting.

An elephant's large ears help the animal to listen for danger. They also act as radiators that lose heat; in this way, elephants keep cool in hot weather.

The elephant has thick legs to support its heavy body and a thick skin to protect it from attacks by enemies.

African elephants

There are two kinds of elephant: the African elephant and the Asian elephant. African elephants are larger and have longer tusks and ears than Asian elephants. But both kinds live in herds that roam among trees eating leaves, branches and fruits.

From time to time, a female in the herd has a baby. A baby elephant is no taller than a big dog. Its mother, or an aunt, protects it from being trampled by the herd and from such enemies as lions or tigers.

Full-grown elephants are so strong that they have no enemies except man. For many hundreds of years people have captured Asian elephants. They tame the animals and train them to lift loads and perform circus acts. African elephants are usually too fierce to be tamed and trusted.

In recent times, hunters have killed thousands of African elephants for their ivory tusks. African countries have set aside lands where guards try to protect the elephant from hunters. But the killing still goes on. Some people fear that one day there will be no more African elephants.

HOW BIG?
An elephant keeps on growing for a long time. Here you can see the sizes of an elephant at 1, 2, 3, 6, 15, 27 and 40 years of age.

40 years
27 years
15 years
6 years
3 years
2 years
1 year

An Indian elephant uses its trunk and tusks to lift timbers nearly four times as heavy as a man. But elephants grow strong enough for such work only by the age of nineteen or so. They are intelligent and learn to stack the loads they move.

Indian elephant

hyrax

The elephant's closest living relative is the hyrax. This little mammal is no longer than a rabbit. Hyraxes have cushioned feet with hooves. Some kinds climb trees. Others scamper up rocks. All live in groups. Their homes are Africa and South-West Asia.

105

gorillas

chimpanzee

Gorillas are the largest apes. A male can stand as tall as a man and weigh three times as much. Gorillas live in family groups. They mainly feed on fruit, leaves and roots. The chimpanzee is the most intelligent of all apes. Chimpanzees can learn to use tools and solve simple problems. Mothers look after their young for several years.

Most monkeys can climb up to thinner, higher branches than the great apes. The colobus monkey and the Diana monkey eat fruit in the tree tops.

colobus monkey

Diana monkey

mandrill

Monkeys and Apes

Monkeys and apes are man's closest living relatives.

Monkeys and apes have man-like arms, legs, hands and feet. On each hand they have a thumb opposite their fingers. This helps them to grip branches and other objects. Apes and monkeys have forward-facing eyes. These help them to judge distance as they spring from branch to branch. Most apes and monkeys are expert climbers. They find food and safety in the trees of the world's warm forests.

Monkeys are smaller than apes and most monkeys have a tail. The tail helps them keep their balance. In many cases, it also acts as an extra hand; a monkey uses its tail to grip a branch when it swings from tree to tree.

Apes do not have tails and they are brainier than monkeys. There are four kinds of ape: gibbons and orangutans in Asia; gorillas and chimpanzees in Africa.

106

The orangutan is a large, red-haired ape from Borneo and Sumatra. Its name means man of the woods.

orangutan

monkey-eating eagle

Some eagles, especially the monkey-eating eagle, are among a monkey's worst enemies.

termite nest

chimpanzee

Chimpanzees can use simple tools. This one is using a twig to dig out termites from their mud nest. The chimpanzee then eats the insects.

The gibbon (right) is smaller than the orangutan, chimpanzee and gorilla. Gibbons have very long arms. A gibbon can swing from branch to branch through the trees without using its feet. Gibbons also have very long, flexible fingers. These help them to grab branches and cling on. The picture shows how the hand of a gibbon compares with an orangutan's hand. Both hands are quite like a human hand. Gibbons live in the hot wet forests of South-East Asia. They make whooping calls to signal to each other. Without these signals they would lose contact among the thick, leafy branches.

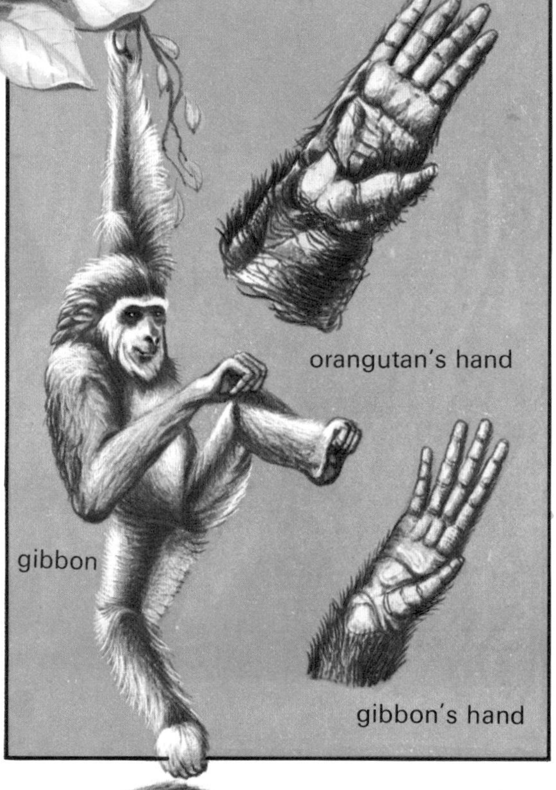

orangutan's hand

gibbon

gibbon's hand

baboons

Not all monkeys are at home in trees. The mandrills and their cousins, the baboons, spend most of their life on the ground. These monkeys live in groups in high, rocky parts of Africa and south-west Asia. Like all monkeys and apes, baboons spend a lot of time grooming. They pick through each other's fur removing dirt, fleas and dry skin. Grooming is friendly and helps keep a group together.

The Meat Eaters

Certain animals eat plants. Other animals hunt the plant-eaters. Sometimes a hunter cannot finish off a large meal. The remains form food for creatures known as scavengers.

True hunting animals kill to obtain all the food they eat. True scavengers eat only what they find already dead. But many animals are a mixture: part scavenger and part hunter.

The puma or mountain lion is a wasteful hunter. It usually kills far more than it needs for food. The dead flesh, or carrion, that is left over often provides a meal for the coyote. Yet coyotes, related to dogs, are also hunters. They catch rabbits, mice and voles.

Hyenas often eat a lion's leftovers. But a pack of hyenas can turn hunter and kill a wildebeeste or zebra. Sometimes, though, a lion drives hyenas from a creature they have killed. Then, the lion acts as scavenger and eats the hyenas' leftovers.

puma

Right: The puma is a big cat found in North and South America. In North America people also know it as the cougar, catamount and mountain lion. This powerful hunter feeds mainly on deer. It kills at night by creeping up on its prey, then pouncing. A puma also eats carrion.

Below: Wolves hold a caribou at bay. These hunters singled out one deer from a herd. They then chased the animal until it was too tired to run any further. After the wolves have killed and eaten the caribou, some meat may remain for crows to scavenge. But wolves do not always kill enough creatures to keep themselves well fed. So, like pumas, they sometimes eat carrion.

Andean condor

The Andean condor is a huge vulture with a wingspan twice the width of a small car. This scavenger lives in the Andes Mountains of South America. It soars high among the mountain peaks and gazes down in search of carrion.

vultures

Dingoes are the wild dogs of Australia. In a group, dingoes attack a cow, a sheep or a wallaby (a small kangaroo). But sometimes a kangaroo kicks and kills a dingo. That way, the hunter becomes a victim and a carrion meal for hungry scavengers.

Vultures look like eagles. But unlike eagles, these birds only eat carrion, dead flesh. A vulture can spot a corpse from high in the sky. As it glides down, other vultures follow. Soon a whole flock gathers round the body and starts tearing it to pieces.

A coyote howls in a lonely, moonlit desert. Today, coyotes also visit towns. There they scavenge scraps of food from people's garbage cans.

coyote

The spotted hyena has strong jaws that can crush the bones of large dead beasts. The hyena eats the soft marrow inside the bones.

hyena

Animals of the Savannas

Vast grasslands, or savannas, stretch across Africa. These savannas, with tall grasses and scattered trees, are home for many animals.

Huge herds of grazing mammals roam the savannas. A mixed herd of zebras and gnus can contain ten thousand animals. Gnus, or wildebeeste, are a kind of big antelope. Mammals that eat grass or feed off shrubs and trees wander slowly across the plains in search of food. The largest herds are found in moist grasslands where there are plenty of pools and rivers for drinking. Zebras, gnus and waterbucks graze in these wetter areas. Small herds of mammals that can last a long time without water are at home in the dry savanna. Some of the small, dainty antelopes called gazelles live in places where rain is rare.

Several plant-eating mammals can share the same feeding area because they do not all eat the same food. Gazelles like tender young grass shoots. Zebras prefer longer, tougher grass. A tiny antelope,

Below: Giraffes and other mammals at a waterhole. A giraffe's long neck and legs are well designed to help it feed off high branches. But when it wants to drink, this animal must spread its front legs awkwardly.

giraffe

rhinoceros

dik-di

the dik-dik, eats low-growing twigs and leaves off trees. The gerenuk is taller and has a long neck. This antelope stands on its hind legs to nibble high branches. The long-necked giraffe can eat leaves that grow too high for any other animal to reach.

Big plant-eating mammals have many enemies. Lions, leopards, cheetahs, hyenas and wild dogs stalk the herds, waiting to kill.

The hunted creatures have some protection. Their colours help to hide them among tall grasses. Also, they can run fast on toes tipped with strong, hard hooves. If an animal hunter chases the grazers, they gallop away quickly. Only old, ill or weak animals are likely to be caught. The fastest and strongest escape.

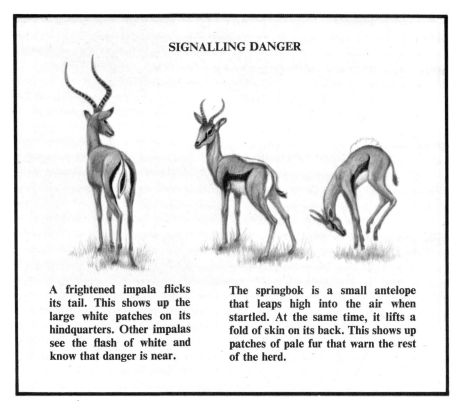

SIGNALLING DANGER

A frightened impala flicks its tail. This shows up the large white patches on its hindquarters. Other impalas see the flash of white and know that danger is near.

The springbok is a small antelope that leaps high into the air when startled. At the same time, it lifts a fold of skin on its back. This shows up patches of pale fur that warn the rest of the herd.

Below: A cheetah chases three impalas while zebras look on unconcerned. Impalas are antelopes that do not have to drink. Eating damp grass gives them enough moisture.

zebra

impala

cheetah

111

Prairie Animals

Once upon a time, many different creatures lived on the prairies, or grassy plains, of North America. Today, most of the prairies are farmlands and only a few wild animals survive.

Many prairie animals eat plants. The largest grazing creature is the bison. A male can weigh a ton. As many as sixty million bison once roamed the plains. By 1900, hunters had killed them all except for two small herds. Now, there are about 12,000 bison living in animal reserves. Early settlers also shot millions of pronghorn. But these speedy animals are still quite numerous.

Prairie dogs also eat plants. These small mammals are rodents. They live in large groups in underground towns made up of burrows. Each family has its own network of tunnels. Another prairie rodent is the pocket gopher. This mole-like creature lives in burrows just below the surface. It grabs plants by the roots and pulls them underground to eat them. Jack rabbits also live on the prairies.

Skunks and American badgers eat small rodents. So do snakes, hawks and eagles. But the chief prairie hunter is the coyote, or prairie wolf. Two coyotes can outwit and kill a pronghorn.

The golden eagle nests among the high mountain peaks, but it soars over the western plains in search of food. from time to time it swoops down to seize prey such as prairie dogs and other small mammals.

golden eagle

Below: Part of a prairie dog town, showing a burrow and its nest chamber cut open. The cone-shaped mound at the entrance comes from soil dug out of the burrow or scraped from the surface. The mound makes a useful lookout post. Also, if rain falls heavily, the water runs off the sides of the mound and not down the burrow.

prairie dog

skunk

Left: Skunks feed on insects, eggs, plants, mice and birds. When frightened, skunks squirt stinking liquid from a gland near the tail. Skunks and American badgers (below) live in many parts of North America.

American badger

112

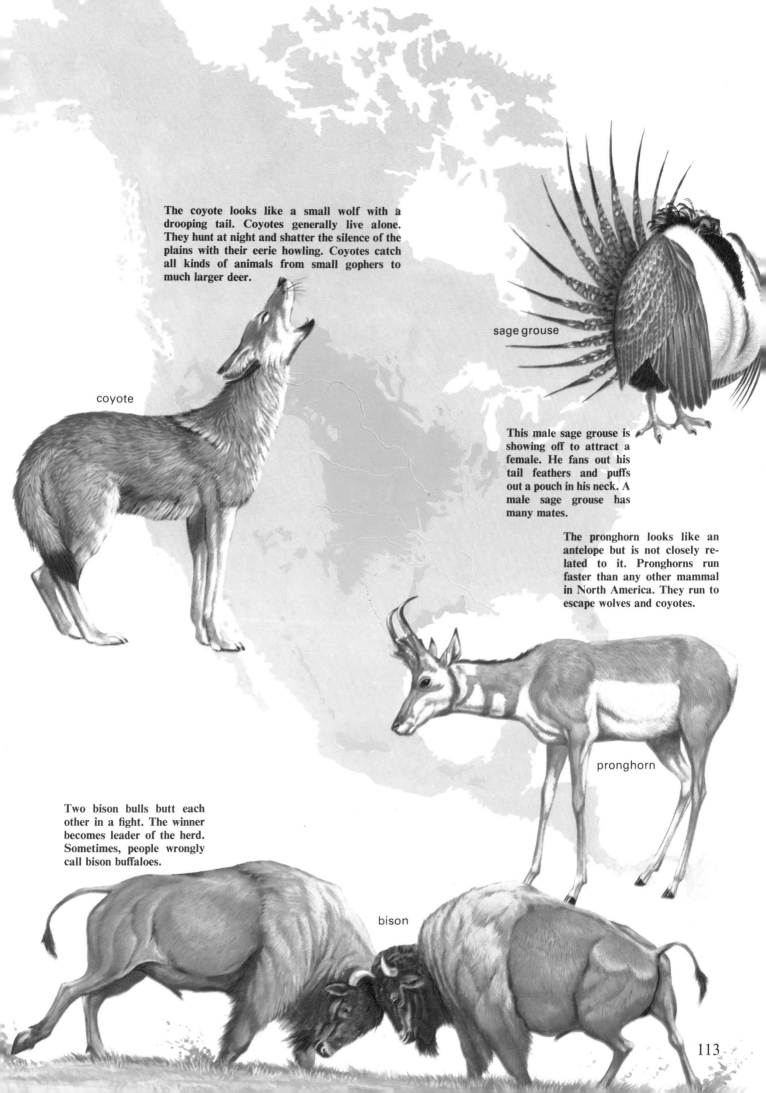

The coyote looks like a small wolf with a drooping tail. Coyotes generally live alone. They hunt at night and shatter the silence of the plains with their eerie howling. Coyotes catch all kinds of animals from small gophers to much larger deer.

coyote

sage grouse

This male sage grouse is showing off to attract a female. He fans out his tail feathers and puffs out a pouch in his neck. A male sage grouse has many mates.

The pronghorn looks like an antelope but is not closely related to it. Pronghorns run faster than any other mammal in North America. They run to escape wolves and coyotes.

pronghorn

Two bison bulls butt each other in a fight. The winner becomes leader of the herd. Sometimes, people wrongly call bison buffaloes.

bison

Animals of South America

horned frog

The wide grasslands of South America shelter many strange creatures.

The grasslands are shown in orange on the map. There are warm grasslands, called campos, in the north and middle and cool grasslands, called pampas, in the south.

Giant anteaters and armadillos live on the campos. Anteaters eat termites, and armadillos eat insects and snakes. The fierce jaguar attacks both animals. Horned frogs feed on small mammals such as mice.

The cool pampas make ideal grazing lands. Deer and guanacos, members of the camel family, graze on the pampas. So do rodents like the mara, plains guinea pig, tuco-tuco and vizcacha. Such rodents, plus birds and reptiles, are hunted by pampas foxes and maned wolves.

Below: A nine-banded armadillo. This species lives on the hot campos grasslands. Armadillos are the only mammals protected by a shell. This is made of horny plates arranged in bands. If attacked, an armadillo rolls up into an armoured ball.

armadillo

These four rodents live on the pampas. The mara has long legs and runs fast if alarmed. The plains guinea-pig hides beneath a tuft of grass. The tuco-tuco and vizcacha hide in burrows by day. At night they come out, unseen by enemies, and graze.

plains guinea-pig

vizcacha

tuco-tuco

mara

Below: The giant anteater is at home on grassy plains as well as in the forests. This mammal walks on its knuckles.

anteater

Animals of Australia

Some of the world's most fascinating animals live on the hot, dry plains of Australia.

Very dry grasslands cover a large part of Australia; they are coloured orange on the map. Kangaroos and their small cousins, wallabies, are the chief grazing animals found on these grasslands. Like many other grass-eating animals, kangaroos live and feed in herds. But unlike other animals, they hop away from danger instead of running. A red kangaroo's strong back legs speed the animal along at nearly fifty kilometres (thirty miles) an hour. One leap can cover seven metres (twenty-three feet).

Rat kangaroos belong to the kangaroo family but look like rats. Some live in burrows, others in busy undergrowth; they build nests of grass and twigs. Another strange Australian mammal is the spiny anteater or echidna. It has a small body, covered in spines, and sharp claws which it uses for digging. The echidna spends the day hidden in hollows, but comes out at night to feed on insects.

The red kangaroo, up to two metres tall, is the world's largest marsupial.

red kangaroo

goanna

The goanna is a big lizard found in very dry grasslands. It climbs up trees to hide.

budgerigar

Budgerigars are small members of the parrot family. Wild ones live in huge flocks on the grasslands.

spiny anteater

Only two mammals lay eggs. The spiny anteater is one of them. The egg hatches in a pouch on the female's belly.

The rat kangaroo is one of Australia's rodent-like marsupials.

rat kangaroo

The wombat is a large marsupial burrower. It digs tunnels up to 15 metres long. Its pouch faces backward so that soil cannot get inside and harm the baby.

wombat

rabbit

Rabbits came to Australia with European settlers. They rapidly multiplied and soon millions of them were destroying pastures.

115

Rivers, Lakes and Swamps

Freshwater swamps, lakes and rivers have their own animal life. Most of the large creatures that belong in freshwater, live in warm parts of the world.

Earth's freshwater giant is the hippopotamus, from Africa. The hippopotamus is one of the heaviest land animals in the world. It spends most of the day in water, with just its eyes and nostrils showing. It swims well and walks on its toes on the river bed. At night, the hippo comes ashore to feed on grass.

Other large freshwater animals include crocodiles and alligators. These reptiles lie on the river bank to get warm. To cool off, they plunge into water or hold open their jaws so that body heat can escape.

kingfisher

pelican

hippopotamus

The Nile crocodile lives in Africa. It can grow twice as long as a man. This crocodile eats fish but at times it will grab and kill animals as big as zebras, and even people. Crocodiles also live in warm parts of Asia, Australia and the Americas.

Crocodiles have relations in the Americas. These are the alligators and caimans. They look very much like crocodiles except that their fourth tooth in their lower jaw is hidden and that their snouts are broader. American alligators can be as large and dangerous as crocodiles. They live in swamps in the south-eastern USA. Females hide their eggs in heaps of leaves and let the sun hatch them. But they often stay nearby to guard their nests.

Caimans, closely related to alligators, live in Central and South America. The largest is the black caiman from the Amazon region of Brazil.

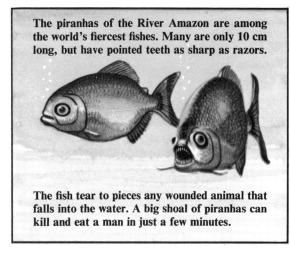

The piranhas of the River Amazon are among the world's fiercest fishes. Many are only 10 cm long, but have pointed teeth as sharp as razors.

The fish tear to pieces any wounded animal that falls into the water. A big shoal of piranhas can kill and eat a man in just a few minutes.

The scene below shows creatures sharing a swamp with an American alligator. The tree frog eats insects. The green snake catches fish. The poisonous cottonmouth takes frogs, turtles and sometimes baby alligators. The turtle feeds on plants and snails. The raccoon likes crayfish.

gharial

The gharial, or gavial, is a large relative of crocodiles and alligators. This reptile has very long, thin jaws. Males have a balloon-like flap on the tip of the snout. The gavial uses its long jaws to seize fish. Unlike crocodiles, gavials are not dangerous to man. They live in big rivers in India.

spectacled caiman

The spectacled caiman gets its name from bony ridges on its head. They look like spectacles. This caiman lives in rivers from Mexico south to Argentina. All caimans have a different kind of nose and different belly scales from alligators.

raccoon

green tree frog

green snake

cottonmouth

American alligator

painted turtle

117

Jungle Creatures

Thick rain forest spreads over the hot regions of the world. The creatures shown here all come from the steamy jungles of South America.

Life is generally easier in the hot, wet forests of the tropics than in the cold forests of the north. In the tropical forests of South America, Africa and South-East Asia, mammals do not need to fight the cold. In fact, keeping cool can even be a problem. So, many mammals in the tropics grow shorter coats than those in cold lands.

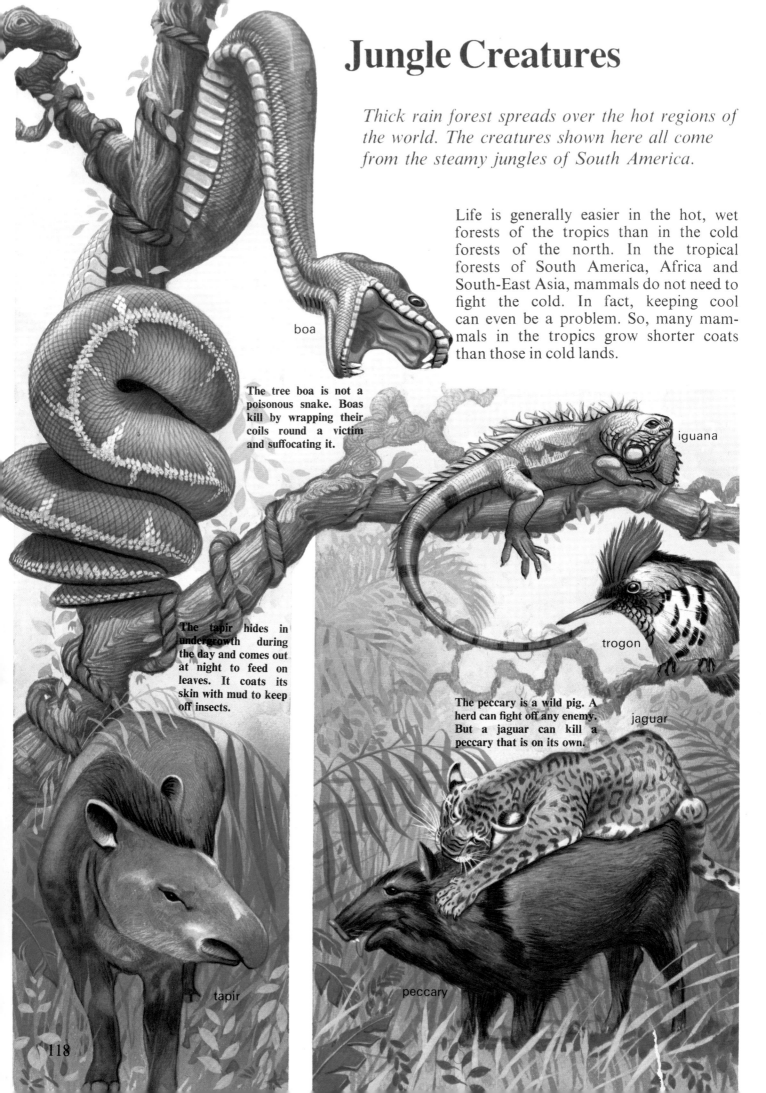

boa

The tree boa is not a poisonous snake. Boas kill by wrapping their coils round a victim and suffocating it.

iguana

trogon

jaguar

The tapir hides in undergrowth during the day and comes out at night to feed on leaves. It coats its skin with mud to keep off insects.

The peccary is a wild pig. A herd can fight off any enemy. But a jaguar can kill a peccary that is on its own.

tapir

peccary

In places where plants grow thickly on the ground, there are not many large mammals. But small creatures like tapirs, peccaries and various rodents can move easily through the undergrowth. Some of these mammals grub about for roots. Others nibble leaves. The giant anteater laps up ants and termites with its sticky tongue. Coatimundis search for insects along cracks in logs.

Treetop mammals are mostly smaller and more agile than the animals below. A woolly monkey uses its long tail to help it balance and to grip branches. Monkeys move very quickly. Sloths move very slowly. These strange mammals hang upside down in trees. Tiny plants grow on sloths and help to camouflage them.

All these mammals have enemies in the forest. Some enemies are mammals, too, like the jaguar. This big cat even attacks animals larger than itself. Some birds and reptiles also feed on mammals. The boa constrictor, for instance, preys on rodents, but the tree boa mainly hunts for birds.

bats

cock-of-the-rock

three-toed sloth

woolly monkey

The three-toed sloth lives in the trees. It can hardly move if put on the ground.

coatimundi

anteater

Giant anteaters tear open termites' nests with their claws. They eat termite eggs, larvae and adults. An anteater is two metres long.

119

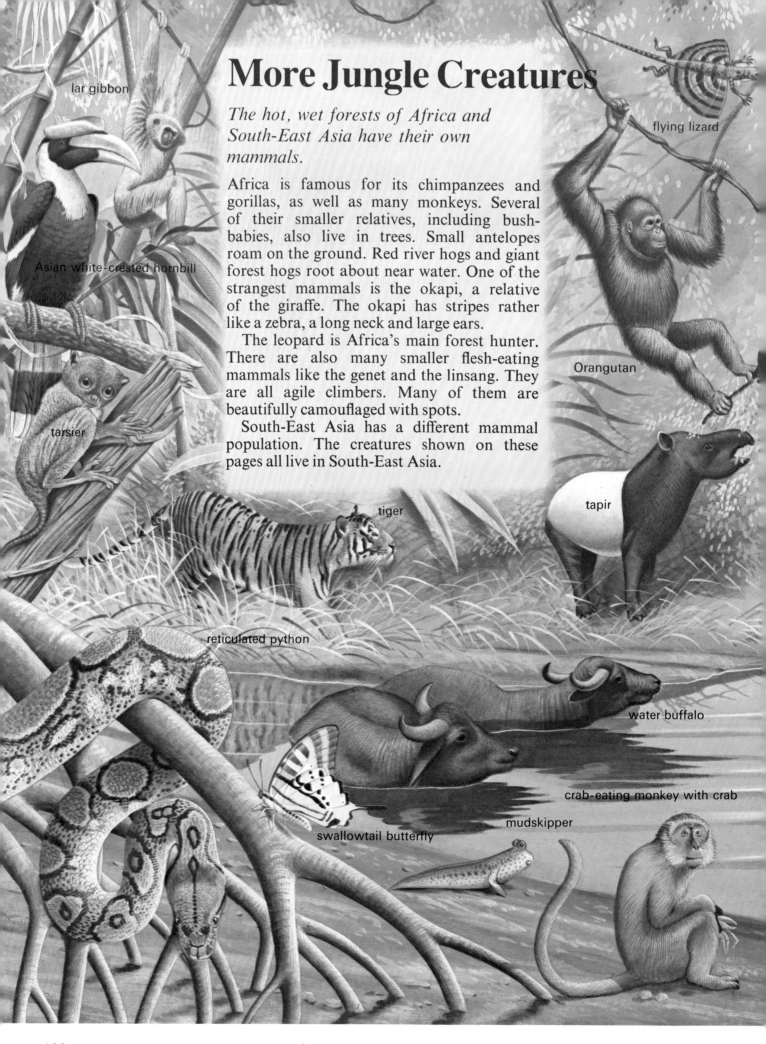

More Jungle Creatures

The hot, wet forests of Africa and South-East Asia have their own mammals.

Africa is famous for its chimpanzees and gorillas, as well as many monkeys. Several of their smaller relatives, including bush-babies, also live in trees. Small antelopes roam on the ground. Red river hogs and giant forest hogs root about near water. One of the strangest mammals is the okapi, a relative of the giraffe. The okapi has stripes rather like a zebra, a long neck and large ears.

The leopard is Africa's main forest hunter. There are also many smaller flesh-eating mammals like the genet and the linsang. They are all agile climbers. Many of them are beautifully camouflaged with spots.

South-East Asia has a different mammal population. The creatures shown on these pages all live in South-East Asia.

lar gibbon

flying lizard

Asian white-crested hornbill

Orangutan

tarsier

tapir

tiger

reticulated python

water buffalo

crab-eating monkey with crab

mudskipper

swallowtail butterfly

Seven of the mammals illustrated here live off the ground among the forest branches. The tree shrew and the tarsier climb about in search of insects. Colugos eat leaves, buds, flowers and fruit. They glide with the help of skin stretched between their limbs. The proboscis monkey also eats leaves. Despite their name, crab-eating monkeys eat more insects and plants than crabs. Orangutans clamber after fruit and birds' eggs. Gibbons swing high to find fruit. A gibbon can leap fifteen metres (forty-nine feet).

Four other mammals in the picture eat plants and live upon the ground. Asian elephants go around in herds; they feed during the cool morning and evening and sleep in between. Water buffaloes, tapirs and rhinoceroses enjoy wallowing in mud. It cools their bodies and keeps insects away. Sadly, hunters have killed many of Asia's rhinos.

Leopards and tigers are the fiercest flesh-eaters. Leopards hide up a tree to ambush monkeys. Tigers go after larger prey. They can even kill water buffaloes.

colugo

proboscis monkey

leopard

Indian elephant

crocodile

green pit viper

Sumatran rhinoceros

tree shrew

Among the Trees

The woods and forests of Europe and North America are rich larders for many different mammals. Some live on the ground, some among shrubs and some high in the trees.

The forest floor is a rich larder for many different creatures, large and small. Rotting leaves form food for earthworms and tiny insects known as springtails. Millipedes and woodlice munch the decaying wood of fallen branches. By day, slugs and snails hide under damp logs. At night, they slide out to feed on juicy plants. Higher up, millions of insects chew the wood and leaves of woodland shrubs and trees.

Plants and tiny creatures provide food for larger animals, especially mammals and birds.

In European woods, voles and shrews scuttle through the fallen leaves. Rabbits graze in open glades. Hedgehogs snuffle as they hunt for slugs and snails. Badgers prowl in search of worms, insects and acorns. Several kinds of deer nibble leaves and shoots. Deer eat bark in winter when the trees have lost their leaves. In some woods, wild boar grub up acorns with their snouts and use their tusks to dig out roots.

Dormice and woodmice climb nimbly among the shrubs over the forest floor. Above them, red or grey squirrels leap from tree to tree, eating buds and seeds.

Many of these mammals fall prey to animal hunters. Weasels scamper to and fro seeking the scent of shrews and voles. Polecats hunt mice and rabbits on the ground. The red fox also eats these animals. Foxes and badgers are the largest hunters in the woodlands now that most wolves and bears have disappeared from deciduous forests.

Wild life in North America's deciduous woodlands is very similar. But North American woods contain more kinds of tree and many animals found there are different from those in Europe. The star-nosed mole burrows through damp, dead leaves searching for worms. Deer mice scamper on the forest floor. The white-tailed deer is the most common deer, but there are also red deer called wapiti. The Virginia opossum is a pouched mammal that lives in trees where it eats fruit and insects. The striped skunk forages for grubs and flesh from dead animals. The raccoon is a cat-sized hunter that flicks frogs and crayfish from the water with its paws. Grey squirrels dart through the trees, chased by martens. Generally, grey foxes are the largest hunters. Wolves, bears and lynxes used to be the biggest flesh-eaters. But hunting and farming have driven them out of the deciduous forests and into the coniferous forests farther north.

Left: Lichens growing on a tree trunk in a wood. Each of these two-in-one plants is made up of an alga and a fungus. Lichens grow only in fairly well-lit woods or forests, and only if the air is very clean. Most lichens grow very slowly. But they provide food for slugs and certain insects. These tiny animals may be eaten in their turn by larger ones.

Right: Springtime in a broadleaved wood in England. A badger and its cub are just about to walk between an oak tree and an ash tree. These are the largest mammals in the picture. But deer are often found in this type of wood. The leaves and plants on the woodland floor hide the runs of voles and shrews. The grey squirrel does not need to hide. This nimble climber escapes most of its enemies by running from branch to branch. All these mammals share their wood with insects and birds like the ones shown. Some insects and birds' eggs become food for badgers.

Life in the North

Life is hard for creatures living in the forests of the far north, especially in winter when it is cold and food is scarce.

Most mammals cannot find food in the great northern forests. This is partly because plant-eating animals generally dislike the resin flavour of the trees that grow there. Also, the soil below these trees is too poor to feed smaller, tastier plants. Then, too, many creatures cannot survive the long, cold winters when snow hides the ground and even less food is available.

For these reasons, fewer mammals live in the northern forests than in forests farther south. Yet, in spite of the food shortage and the bitter winters, some animals do make their home among the cold conifers.

lynx

The lynx belongs to the cat family. This hunter catches hares and grouse by creeping up and then pouncing.

The capercaillie is a large grouse. It lives in the northern forests of Asia and Europe. The lynx is one of its enemies.

The red squirrel mainly eats the seeds from cones. This squirrel buries hoards of cones in summer and eats them during the winter.

red squirrel

capercaillie

pine marten

chipmunk

Chipmunks are small squirrels. They like to eat seeds and berries. Chipmunks store food in their cheek pouches.

The pine marten is a small fierce hunter. It chases squirrels in the trees.

brown bear

Brown bears eat plants and animals. They munch shoots and berries and often steal honey from the nests of wild bees. They also catch rabbits, small deer and fish, like this salmon. A bear kills its prey by hitting it.

salmon

C.J.KING

The mammals of the northern forests have thick fur. This fur stops heat leaking from their bodies and so keeps them warm in winter. Some of them actually feed on the unpleasant-tasting conifers. Others manage to find different sorts of plant foods. The flesh-eating animals hunt the plant-eaters.

Most mammals have their babies in spring. The young grow in summer when there is plenty to eat. The food they eat in autumn helps them to survive the winter. In winter some mammals doze for several weeks. They use less energy, and so need less food, than animals that stay active. However, very few mammals of the cold forests truly hibernate.

The most common mammals in the forests are rodents. Rodents are plant-eaters with long, sharp, front teeth shaped like chisels. The smallest rodents in these cold lands are voles and lemmings. In winter these little animals stay warm and active in snow burrows. High above, tree squirrels feed on bark and pine cones. Flying squirrels glide from tree to tree with a skin parachute stretched between their legs. The largest rodent found in these forests is the beaver. It is the second biggest rodent in the world.

The largest plant-eaters are deer. In winter the caribou of North America and the reindeer of Europe and Asia move south from the Arctic to seek food and safety among the trees. So do the North American moose and the elks of Europe and Asia.

Many plant-eaters in the northern forests are hunted by flesh-eating mammals such as lynxes, bears, wolves and foxes.

A moose feeding in a forest. Moose, from North America, are the world's largest deer. They eat plants that grow in and around forest lakes.

The wildcat looks like a large tame tabby cat. But this animal is larger and much fiercer. Wildcats hunt rabbits and birds in European forests.

This is a beavers' home or 'lodge' cut open to show how it is made. Beavers are master builders. They dam a stream to make a pond. Inside the pond they raise a mound of mud and sticks. Then they chew tunnels into it from underwater. A beaver family lives in a room hidden above the water level. There are several underwater entrance tunnels.

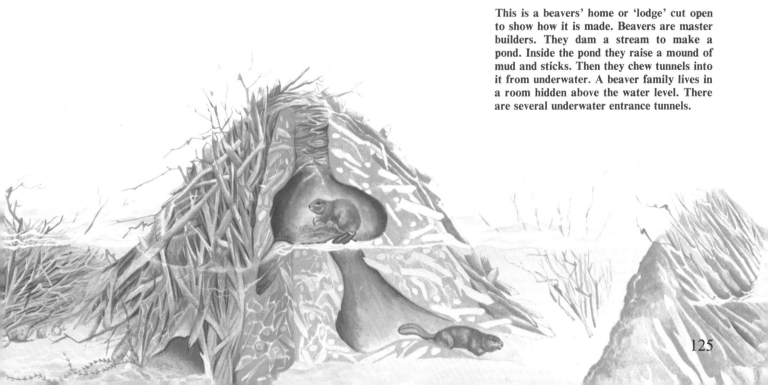

Creatures of the Snow and Ice

Polar lands, at the ends of the Earth, have long, cruel winters. Summers, too, are cool. Yet some animals live here all the year.

Right: Penguins leap from the sea onto an ice shelf in Antarctica. As well as penguins, some other water birds and mammals breed on this icy continent. But most creatures cannot find food on the frozen land and waters.

Life is hard in both the polar regions. But summers are warmer in the Arctic lands of the far north than in Antarctica in the far south. During the Arctic summer, the sun shines most of the time and plants grow quickly. There is plenty of food for small plant-eating mammals like lemmings, voles and Arctic hares. The much larger musk oxen and caribou also find enough to eat. Weasels, stoats and Arctic foxes hunt the small plant-eaters. While wolves and polar bears attack caribou and musk oxen.

In winter, all these creatures must fight the cold. Small mammals hide in burrows under the snow. In their burrows, voles and lemmings

musk ox

The musk ox (left) looks like a small cow, but is a close relation of the sheep. If wolves come near, musk oxen form a ring and face outwards. Musk oxen and caribou (below) paw away snow to reach food plants. The caribou has hollow hairs that help to trap body heat and keep it warm.

caribou

stoat

Below: In winter, voles and lemmings feed on roots and stems under the snow. Big enemies cannot get through to them. But weasels are slim enough to catch them in their burrows.

lemmings

voles

Weasel

126

Adélie penguins

stay wide awake all winter. But Arctic ground squirrels hibernate. No other Arctic mammal does so.

Mammals that are too big to burrow find other ways of beating the cold. Thick fur protects the Arctic hare and Arctic fox. The hare finds food on slopes swept bare of snow by the wind. The Arctic fox hunts Arctic hares. Caribou and wolves escape the cold by trailing south for hundreds of kilometres to reach the shelter of the forests. The hardy musk ox, however, braves the dark, cold winters in the open. Its thick inner fleece and long outer hair shield it from the bitter winds.

Life is very different in Antarctica. Snow and ice cover most of the land throughout the year. Penguins and seals both come ashore to breed. But no large land animals can find a living on this frozen continent.

| fennec fox | red fox | Arctic fox | jackrabbit | hare | Arctic hare |

Above: Foxes and hares with ears of different sizes. Big ears help an animal to lose heat. Small ears help to keep heat in. The Arctic fox and Arctic hare have small ears. These help to keep them warm in the cold Arctic winter. The fennec fox and jackrabbit have long ears. These keep them cool in the hot deserts where they live. The red fox and hare have medium-sized ears. Both animals live in places that are neither very hot nor very cold.

The Arctic hare (below) and the stoat (left) have brown summer coats, but turn white in winter to match the snow. Stoats in their white winter coats are often called ermine.

Polar bears come ashore to eat berries, stranded whales and other foods. But they chiefly hunt on sea ice where they ambush seals.

This large bull walrus is twice as heavy as a cow (female) walrus. Walruses rest and breed on shore, but feed at sea.

Arctic hare

polar bear

walrus

127

At Home in the Desert

A few deserts are too dry and too harsh for any living creatures to survive. But most deserts are home for many kinds of animal.

Desert animals have special ways of surviving the heat and the drought. The different creatures on these pages all manage to live in the hot, dry deserts of south-western USA. Most of them hide underground during the day when it is very hot. But, as the sun goes down, the desert surface quickly cools. Then, in the dark, the animals creep out of their holes to feed.

Kangaroo rats eat seeds scattered by the plants that bloomed last time it rained, perhaps many months ago. The seeds hold all the moisture that their bodies need. Jackrabbits are also plant-eaters that live without drinking. Their long ears lose heat and so help keep them cool by day. Large ears also help warn them if enemies approach. Not all plant-eaters can live off dry foods. Some pack rats feed on juicy cactus fruits.

Many small plant-eaters are attacked by falcons, foxes and snakes. These hunters obtain the water they need from the bodies of their victims. In the same way, grasshopper mice get food and water by eating insects.

Desert animals must somehow stop moisture from their bodies leaking into the air. Most mammals lose moisture in sweat and body waste. Kangaroo rats, for example, do not sweat and they produce dry droppings. Spiders, scorpions, snakes and lizards all have waterproof skins that trap the moisture in their bodies.

cactus

The prairie falcon is an agile flyer with keen eyes. It swoops suddenly to seize small rodents.

prairie falcon

jackrabbit

The jackrabbit is really a hare and not a rabbit. It can run at 70 kilometres per hour.

The horned toad is a desert lizard that eats ants. Spiky scales guard its body.

Kangaroo rats have long back legs and leap about like tiny kangaroos.

horned toad

The desert rattlesnake slithers over sand by making S-shaped sideways movements.

rattlesnake

128

trapdoor spider

THE HIDDEN TRAP

The trapdoor spider is a small tarantula that lives in a tunnel lined with silk. The creature makes a hinged silk lid and covers this with sand to match the ground around it. The spider hides inside its home. When an insect passes, the spider leaps out and pulls it underground.

A RATTLE FOR DANGER

An angry rattlesnake threatens enemies by rattling its tail. Its rattle is made of horny plates that fit loosely together. A new plate is added to the rattle each time the creature sheds its skin.

TRANSPORT IN THE DESERT

Camels can live for days without eating or taking a drink.

Camels' broad feet help them to walk easily over desert sand.

Arabian camel

Bactrian camel

kit fox

The kit fox has large ears that lose heat and so keep it cool.

The tarantula is a large, hairy spider. Its poisonous bite kills insect prey.

The scorpion is a relative of the spider. Its tail carries a poisonous sting.

scorpion

tarantula

grasshopper

The scorpion kills insects with poison squirted from the sting in its tail.

129

The Horse Family

Man has tamed some wild animals. He uses them for food and for work. Animals like sheep, pigs, goats, cows and horses are part of daily life. Especially horses.

Haunch · Croup · Loins · Withers · Back · Shoulder · Crest · Mane · Atlas · Forelock · Cheek Bone · Nostril · Throat · Chin Groove · Jugular Groove · Dock · Flank · Buttock · Thigh · Stifle · Gaskin · Hock · Sheath · Ribs · Abdomen · Chestnut · Knee · Shannon · Cannon Bone · Coronet · Pastern · Fetlock · Hollow of Heel · Heel · Breast · Elbow

This picture shows the different parts of a horse. When people buy a horse, they choose an animal that is well shaped. The eyes and ears should be alert. The back must be straight, not hollow. The legs should be firm and straight. The hindquarters should be strong. These facts also apply to ponies. A pony is just a small horse.

A horse's height is measured in hands from the ground to the top of the withers. A hand is 10 cm (4 inches). Horses are 14 hands 5 cm or more; ponies are under this height.

Since ancient times, horses have helped man in many ways. Horses have taken soldiers into battle. Horses have worked as pack animals carrying heavy loads. Horses have pulled plows, wagons, carriages and even buses. In some parts of the world, horses still do these jobs; but in many places, cars, trucks and tractors have taken over. Yet, even in these countries, there are plenty of horses. Horses for sport, for entertainment and for fun.

There are various reasons why horses are so useful, all connected with how horses are made and how they live. Horses are large and strong enough to carry a person. Horses eat grass, a food that grows in most parts of the world. Horses can run very fast, faster than most other mammals. They gallop quickly because they have long legs and rest their weight on their toes. Each foot ends in one large toe, protected by a huge, hard nail called a hoof.

THE FIRST HORSES

The horse's early ancestors lived 55 million years ago. One was fox-sized *Eohippus* (dawn horse). *Eohippus* ran on slender fingers and toes and fed off soft leaves in the forests.

Mesohippus (middle horse) came later. This beast was as big as a sheepdog. *Mesohippus* had a large middle toe on each foot; its outer toes were much smaller.

Merychippus (cud-chewing horse) was a larger, later, prehistoric horse. Only the big middle toe of each hoof touched the ground. This horse ate grass on the open plains.

Eohippus

Mesohippus

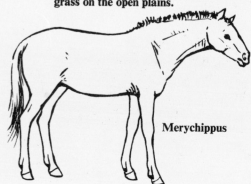

Merychippus

mule

zebra

donkey

onager

The horse's wild relatives are asses and zebras. They all have a single-hoofed toe on each foot.

Wild asses roam hot deserts and dry grasslands in Asia and Africa. One kind is called the onager. Wild asses can go without a drink until their bodies are so dry that they have lost three kilograms (6 lb) for every ten (22 lb) they weighed. Yet, in a few minutes they can drink enough water to make up for all they have lost.

Donkeys are tame asses. They are smaller than horses. In poor countries, people often use donkeys to carry heavy loads. A male donkey and a female horse can be mated to produce a mule. A mule is bigger and stronger than a donkey and can work very hard.

Zebras stand out from other members of the horse family because of their startling stripes. The stripes help to hide them from enemies by breaking up their outline against the trees and grass. Zebras live in small herds in Africa. One kind, the quagga, is now extinct. It had a striped head and neck, a brown body and white legs.

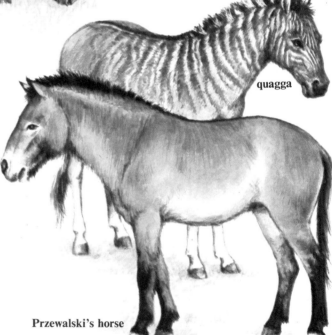

quagga

Przewalski's horse

HOW A HORSE MOVES

A walking horse raises its legs one after another. A walking horse has a four-beat rhythm.

Trotting is faster than walking, and has a two-beat rhythm. A rider learns to 'rise to the trot'.

Cantering is faster still and has a three-beat rhythm.

Galloping has a four-beat rhythm, like walking, but is much faster.

Breeding

The wild ancestors of horses and ponies were mostly small and sandy-coloured. But breeding has produced animals of all sizes and colours.

Each horse colour has a name. Dun horses are yellowish-grey. A chestnut is reddish-brown. A bay is brown with a black mane, tail and legs. Palominos are golden with a pale mane and tail. A grey has mixed black and white hairs. It is born black and goes white with age. Roans have a solid colour with white hairs. Many horses have large patches of white. Piebalds have white patches on black. Skewbalds are any colour except black, with white patches.

Below: Four of the different face markings found on horses. Each type of marking has a name.

blaze

snip

star

white face

dun

brown

strawberry roan

bay

palomino

piebald

chestnut

skewbald

grey

black

132

Icelandic pony

Shetland pony

Appaloosa

Welsh cob

Thoroughbred

Arab

This page shows several breeds of horse and pony. The Shetland pony is one of the smallest breeds in the world; it is only a metre (39 in) high. Like Shetlands, Icelandic ponies are very hardy. One of the world's largest horses is the Shire. This powerful work horse is nearly twice as high as a Shetland pony and can weigh a ton. The Breton, from France, is another strong work horse. Arabs, Welsh cobs and Morgans all make good riding horses. Most racehorses are thoroughbreds or quarter horses. The Lipizzaner, from the famous riding school in Vienna, can be taught to jump and dance. The spotted Appaloosa often appears in spectacular circus acts.

Lipizzaner

Quarter Horse

Morgan

Breton

Shire

133

Out at Work

Years ago, horses did all kinds of work. Today, machines have taken over. Yet, in the country and the city, some jobs are still done by horses.

There are several reasons why horses are still working. They are actually better than machines for some jobs. They breed and multiply. Machines do not. They are cheaper to buy and look after than certain machines. Lastly, horses live on grass; machines use a fuel, like oil. The world has plenty of grass, but it is running out of fuels.

police horse

pit pony

coal cart

The heavy work of pit ponies and Shire horses is now mostly done by machines. Pit ponies were small, sturdy ponies that worked in coal mines. Each one was harnessed to a truck which ran on rails. Miners filled the trucks with coal, then the animal pulled them away. The ponies lived in underground stables for most of the year. They only saw daylight when they came up for a rest on a farm. Shire horses pulled plows and heavy farm wagons.

Fuel-powered taxis, cars and buses have replaced horses for travel in towns and cities. But horses are still seen on the streets: police horses. A mounted policeman can edge his horse sideways towards a crowd to stop it spilling out into the road. Police in cars or on foot cannot control crowds so well.

Shire horses

plow horses

134

Australian stockmen

Above: Australian stockmen rounding up cattle. When they are not working, stockmen take part in rodeos very like the rodeos of North America. They ride bullocks and try buckjumping, an event like bronco busting.

Below: A mustang ridden by a gaucho. This South American cowboy wears a cape called a poncho.

Horses can move fast over ground that is too rough for wheels to run on. Horses can also stop or turn quickly. This means a cowboy on a horse can chase and catch a cow as it gallops, twisting and turning, through bushes, trees and rocks.

Cowboys are not just a legend of the Wild West. Modern cowboys herd sheep and cattle in North and South America, Australia and New Zealand. Australian herdsmen are called stockmen. They ride Australian stock horses, which are a mixture of breeds. South American cowboys are known as gauchos. Some of them ride mustangs. These horses come from the wild herds that roam the plains.

gaucho

135

Running Wild

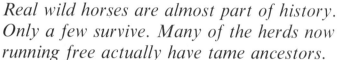

Real wild horses are almost part of history. Only a few survive. Many of the herds now running free actually have tame ancestors.

The wild horses of North America have an interesting story. The continent's true wild horses died out thousands of years ago. Perhaps they were killed by Stone Age hunters. When Spanish explorers reached North America five hundred years ago, they found there were no horses anywhere in the land. In fact, the horses belonging to the Spaniards terrified the natives. The American Indians had never seen such fierce monsters.

As time went by, some of the Spanish horses escaped and became wild. These wild animals bred and increased in numbers; soon, there were big herds of them roaming the plains. Over the years, Red Indians caught and tamed some of these small, sturdy horses. They are now called mustangs. Mustangs look scruffy and can be bad tempered and cunning, but they are also brave and tough.

A herd of mustangs takes fright at a sudden sound. Large herds of these wild horses once roamed the vast plains of North and South America. As the picture shows, mustangs come in a variety of colours. They are independent animals and difficult to tame.

Camargue horses splashing through the shallow marshes of southern France. These small, hardy horses graze on tough grasses. They are sure footed and can jump and swim with ease.

Other countries also have wild horses descended from tame European breeds. South America's wild horses came from Spain. The ones in Australia had British ancestors.

Europe itself has several breeds of horse that are, at least, half wild. One of the best-known is the Camargue horse. These graceful, grey-white animals belong to the Camargue, a vast, empty marshland in southern France. There, they endure cold winter winds and a burning summer sun. From time to time, French cowboys called gardiens round the animals up and break them in. But no one knows just when or how the Camargue first got its grey-white horses. Some experts think the horses go back to Stone Age times. This is almost certainly true of Britain's half-wild Exmoor ponies and other mountain breeds.

The only true wild horse left in the world today is Przewalski's horse. This small, yellowish-brown creature lives on the desolate plains of Mongolia in central Asia.

Przewalski's horse is a small animal that looks like a yellow-brown donkey. A century ago, the Russian explorer, Nicolai Przewalski, discovered it living wild on the plains of Mongolia. But Stone Age hunters knew about this horse. They drew pictures of it on the walls of their caves.

137

Index